Q & A with Dr. K

Why it is so

Paul,

This is my second favorite Polish Man.

Hope you enjoy

Take care and here's good thoughts for a Speedy Recovery.

Love & Thanks

Jan

Q & A with Dr. K

Why it is so

Karl Kruszelnicki

HarperCollinsPublishers

HarperCollins*Publishers*

First published in Australia in 2001
by HarperCollins*Publishers* Pty Limited
ABN 36 009 913 517
A member of the HarperCollins*Publishers* (Australia) Pty Limited Group
www.harpercollins.com.au

HarperCollins*Publishers*
25 Ryde Road, Pymble, Sydney, NSW 2073, Australia
31 View Road, Glenfield, Auckland 10, New Zealand
77–85 Fulham Palace Road, London, W6 8JB, United Kingdom
2 Bloor Street East, 20th floor, Toronto, Ontario M4W 1A8, Canada
10 East 53rd Street, New York NY 10022, USA

National Library of Australia Cataloguing-in-Publication data:

Kruszelnicki, Karl 1948– .
Q & A with Dr K.
Bibliography.
ISBN 0 7322 5855 3.
1. Science-Popular works. 2.Questions and answers. I. Title.
500

Cover photograph: Karin Catt
Cover and internal design: Maxco Creative Media
Internal layout: Helen Gibson
Illustrations: Adam Yazxhi, Maxco Creative Media
Printed and bound in Australia by McPherson's Printing Group on 80gsm Econoprint

10 9 8 08

THANKS

First, I thank the architects, research scientists, engineers, veterinarians and medical doctors who helped me — Garry Cross, Ralph Burley and Julian Cox (Mike the Headless Chicken), David Willey (Fraudulent Firewalking), Michael Halmagyi and Ian Curthoys (Room Spins When Drunk), John Prescott (Chilli — Burn Baby Burn), Alexander Kira and Tony Wren (Bathroom Queues Blues) and Rachel Herz (Smell and Memory). The *Real Science Experiments* in this book depended on Luke Tennent (for doing probably the world's first Fartograph) and Steve Manos (for statistically teasing out the results from the enormous Belly Button Fluff (BBF) Survey, and for making electron microscope photographs of the Fluff). Adam Spencer came up with some very clever lines. I would also like to thank the phenomenal photographic flair of Karin Catt (we love you), who always manages to squeeze me into her super-hectic schedule. And of course I thank Teri Thomas for writing the original and best Mike the Headless Chicken book, and for giving us lots of leads.

Second, I thank my family for reading the stories and improving them — and for still loving me while I was writing.

Third, I apologise to the staff at HarperCollins Australia for increasing their work load by being five months late in delivering the manuscript — the ever-obliging Emma Kelso (elegant and eloquent editor extraordinaire), associate publisher Alison "Al We Love You and Will Never be Late Again" Urquhart, and Adam "The Only Man with the Last Three Letters of the Alphabet in his Name" Yazxhi, who illustrated so fabulously and quickly. If this Big Rush has led to any mistakes — it's all my fault!

Fourth, I thank the webmasters of our prize-winning science web site (www.abc.net.au/science/k2) at The Lab — Ian Allen and Kylie Andrews. They set up the BBF survey on the web. Frankie "Leftie" Lee did the publicity for the Belly Button Survey.

Fifth, I thank my agent team — Rosemary Creswell, and her colleague Annette "Still Undiscovered Singing Talent" Hughes.

Sixth, I thank Caroline Pegram — especially her leadership skills and calmness under pressure, combined with her ability to find anything anywhere on the planet. Caroline has also set up her own personal International Network of Professor Friends — who gave many answers to my questions.

And finally, I dedicate this book to my parents, who taught me so much.

CONTENTS

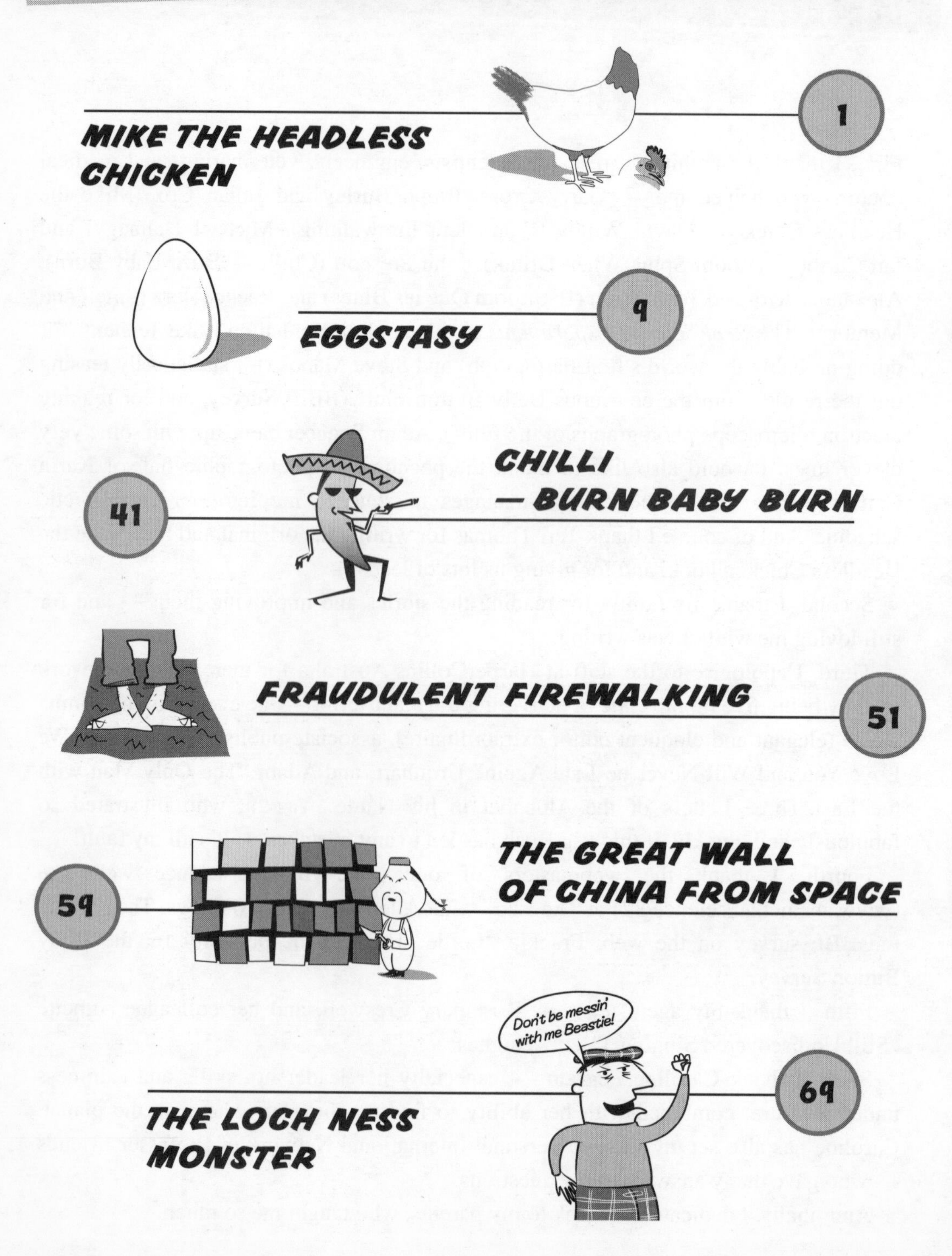

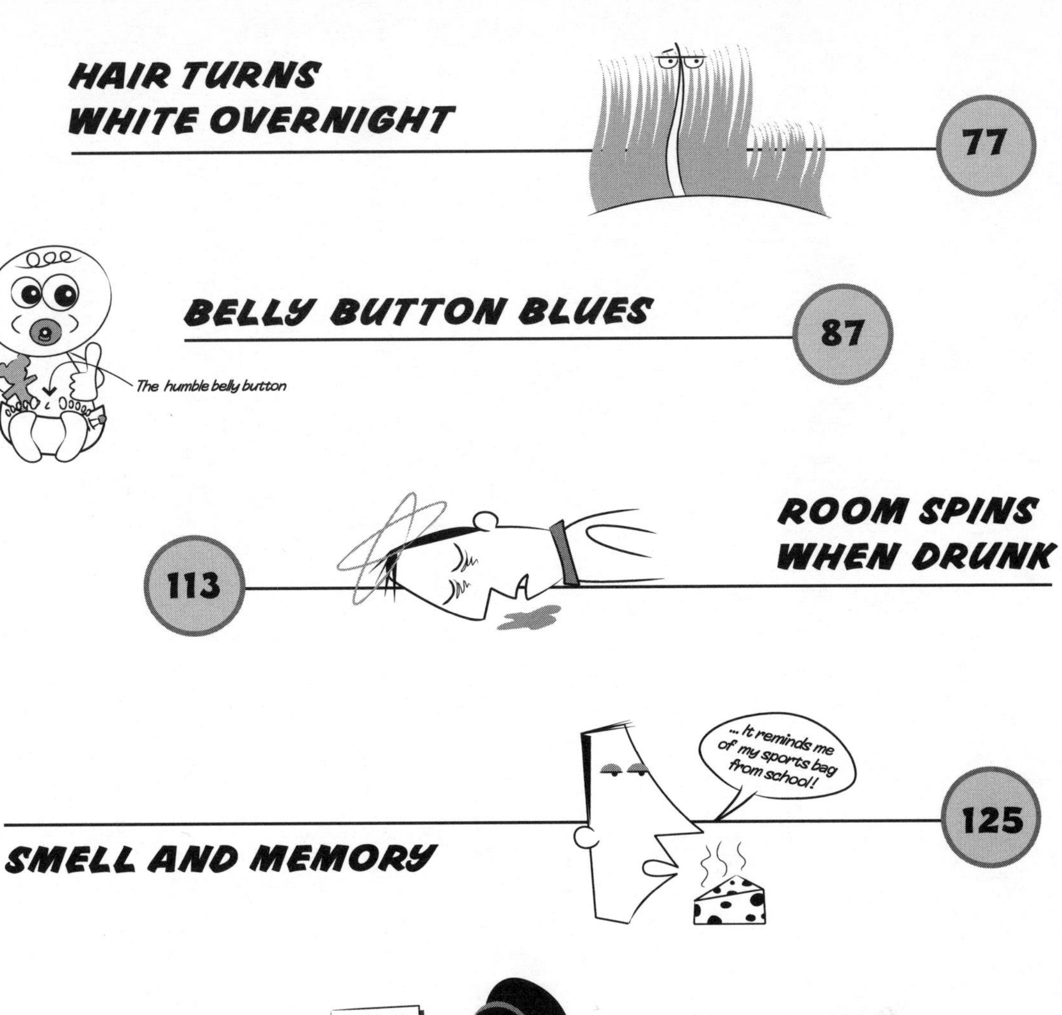
The humble belly button
... It reminds me of my sports bag from school!

< LADIES
PLEASE TAKE A NUMBER
number 248

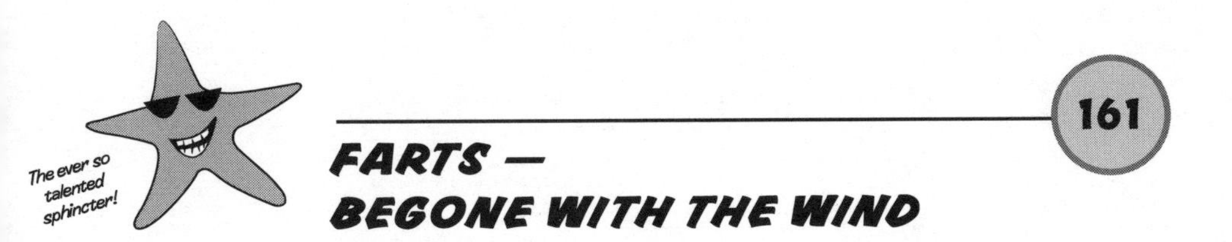
The ever so talented sphincter!

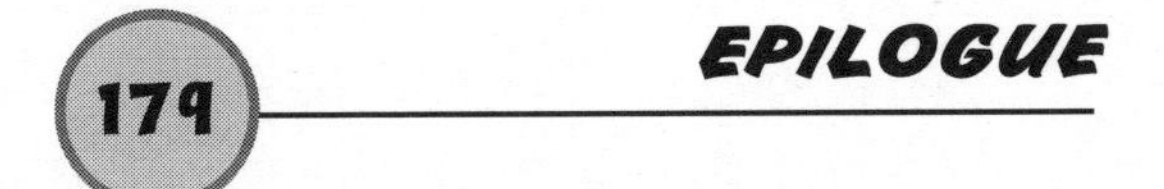

Mike THE HEADLESS CHICKEN

There was a chicken that had no haid,
How was it that this chicken weren't daid?

"*Running around like a chook with its head cut off*' is a very typically Australian description. If you ask people who have chopped heads off chooks, you'll soon find some who'll agree that they've seen a headless chicken running around the yard for up to 30 minutes before dying. This is strange, because a human dies within seconds when decapitated.

But it gets even stranger. Mike the Headless Chicken from Fruita, Colorado holds the World Record for Headless Chook Survival. He kept walking for a year and a half after his head was chopped off.

THE "LUCKY" CUT

Fruita is about 320 km west of Denver, Colorado. Back on 10 September 1945, Lloyd and Clara Olsen were slaughtering some 45 chickens. Lloyd decapitated the chickens, and tossed the bodies and the heads into two separate piles. Clara plucked and cleaned the bodies.

Lloyd and Clara had each lived through the Depression of the 1930s, so they didn't like to waste anything. Furthermore, World War II had just recently ended. People were still suffering shortages, and consumer goods were rationed. So when Lloyd cut the chickens' heads off, he made his cut as high as possible on the neck, to keep as much meat as he could.

In the case of one Wyandotte rooster, Lloyd accidentally made a rather unusual swipe with his axe. This cut was so high along the neck that the rooster's body was left with most of the cerebellum, the medulla oblongata and one ear.

ODES TO MIKE

Today Mike has been given his full and deserved recognition.

Teri Thomas, daughter of the editor of the *Fruita Times*, examined all the files and wrote *The Official Mike The Headless Chicken Book — a 1940s Tale of Two Men and a Chicken.*

Fruita has an annual Mike the Headless Chicken Festival every May. The activities include the 5-km Run Like a Headless Chicken Race, and Chicken Bingo (your number is allocated according to where chicken droppings land on a numbered grid).

Lyle Nichols, who grew up in Fruita, has made a 136-kg rusty metal replica of Mike from various pieces of discarded farm equipment.

And Greg Finch, designer of the Official Mike the Headless Chicken Cybercoup Website (www.mikethe headlesschicken.org) said that "Mike is the greatest — he's empowered me with the knowledge that I can lead a full life after I have lost my mind."

Once the chopping was all done, Lloyd joined Clara in preparing the chicken bodies. As they worked their way down through the pile, they found the headless Wyandotte rooster still alive. They left him alone until they had finished cleaning the rest of the chickens — and he was still breathing. Lloyd decided to leave Mike the Rooster overnight to see if he would survive — and he did.

MIKE GETS FAMOUS

Lloyd started feeding Mike. He squirted water, cod liver oil and ground grain down his food pipe (oesophagus) with an eyedropper. He also added the occasional little pieces of rock. Mike needed rocks, because chickens don't have teeth; they grind their food in their gizzards. A gizzard is a hollow muscular organ, right at the bottom of the long neck and just inside the body. Chickens deliberately eat small pieces of gravel to speed up the grinding process. Mike thrived on his new diet, and started putting on weight.

Word of Mike the Headless Chicken soon spread, and Hope Wade, a promoter, contacted the Olsens. He convinced them to take Mike to the University of Utah in Salt Lake City for scrutiny. (Unfortunately, however, all the records of that examination have been lost due to the lapse of time.)

Life magazine heard about Mike while he was still in Salt Lake City, and wrote a story about him for the 22 October 1945 issue (the one with the Ohio State star halfback on the cover). Mike then made it into *Time* magazine on October 29. However, the famous photo of Mike standing proudly next to his head is not accurate. It is correct that he does not have a head on his neck. But the head on the ground next to his feet is not his head. His head was eaten by the farm cats. The head in the photo is from another random chicken.

WHY MIKE SURVIVED — IS IT SCIENTIFICALLY POSSIBLE?

According to Professor Garry Cross, from the Faculty of Veterinary Science at the University of Sydney, it's quite possible for a chicken to survive without a head — if the cut is very high on the neck.

This is because the chicken skull is mostly full of eyeballs — quite different from human skulls, which are mostly full of brains.

Chickens have a very large eye for their body size. The small, beady circle of eye that we see is only a very small part of the chicken's very big eyeball. So chopping off the chicken's head removes the eyes, but only a small amount of brain.

MIKE'S SURVIVAL — BLEEDING

Lloyd Olsen said that the neck looked as though it had been painted with a red paintbrush, but that it wasn't leaking blood. Why didn't Mike bleed to death as soon as he was decapitated?

There are four reasons.

What has 2 legs, feathers and no head?

Back on 10 September 1945, Lloyd and Clara Olsen were slaughtering chooks. After cutting the heads off quite a few, they realised that one rooster was still alive. They left the headless one overnight, and the next morning he was still alive. Even though Mike took a lickin', he kept on tickin'.

OTHER HEADLESS CHICKENS

It seems as though Mike was not the only headless chicken to survive. The *Guinness Book of Records* of 1980 (the children's version) discusses Biddy, a black Minorca hen. She was beheaded in 1904 by a hotel owner in Michigan. She was fed successfully via a syringe for 17 days. Unfortunately, the skin on her neck closed over her windpipe, and she stopped breathing.

Other headless chickens who have lived for days or weeks have been reported in Coal City in Illinois, Fresno in California, Mobile in Alabama, Middle Port in Ohio and Walnut Grove in Missouri.

First, Lloyd's axe cut Mike's neck so high that it missed the big blood vessels, which would have emptied all Mike's blood within a few minutes. Instead, the axe cut through only small blood vessels.

Second, the axe blade was not super-clean. If you accidentally get cut by a sterile scalpel, you will bleed for a long time. But a "dirty" blade stimulates your natural clotting mechanism into making blood clots, which then plug up the holes from which blood could leak.

Third, the axe blade would have given a ragged cut, which would have caused "compressive separation". This would close the small capillaries, releasing chemicals called "tissue factors", which would also enhance the clotting reaction.

Fourth, when Mike's head was chopped off, he almost certainly went into shock. His heart rate and blood pressure would have plummeted. The blood wasn't pushing so hard on the clots, so they were more likely to remain stable.

The near-fatal axe blow to Mike's head was very high ... so high, in fact, it left Mike with one ear. The axe also missed the big blood vessels that would have caused Mike to bleed to death.

It was very important that Mike had at least one ear. Apart from hearing, an ear provides balance and tells you which way is up. Also, the part of the brain that keeps the heart beating and the lungs breathing was left intact in the back part of Mike's brain.

MIKE'S SURVIVAL — LUNGS AND HEART

Lucky Mike lived on for many reasons. One was that the really essential part of the chicken brain, the medulla oblongata, was left safely attached to his body.

In humans the cerebrum, the front part of the brain, deals with abstract concepts like poetry, income tax, music and nuclear weapons. Our human cerebrum is very large, and has an extensive surface area because it is criss-crossed by mini-ravines. It needs lots of surface area, because a huge amount of information processing happens there.

In chickens, the cerebrum is much smaller and has hardly any mini-ravines. So a chicken can probably have only two thoughts at a time — and if it wanted to think a third thought, it would probably have to forget the first one. (If it realised that it had to forget the first thought in order to have a third thought, that would, in itself, constitute a fourth thought — and the chicken would be very confused.) The cerebrum is not essential for chicken life.

However, the medulla oblongata is absolutely essential for animal life. It's below your cerebrum, and is part of the brain stem. The medulla oblongata has two pacemakers that control the lungs and heart. If these pacemakers stop, your lungs stop breathing and your heart stops beating.

HEAD WITHOUT A BODY

In 1971, Dr Robert White from the Case Western Reserve University in Cleveland, Ohio, claimed that he and his team of five surgeons had been able to keep the severed heads of Rhesus monkeys alive for as long as 36 hours.

In 1988, Chet Fleming, a molecular biologist from St Louis, was granted US patent 4666425. It discusses a hypothetical device that would keep a severed head alive by feeding it all the oxygen, food and fluids that it needed. It would do this via a series of pipes.

YEAR OF THE ROOSTER

Mike was born around May, 1945. In the Chinese Calendar, that puts him in the Year of the Rooster, which ran from 13 February 1945 to 1 February 1946.

The Rooster is born under the sign of candour. He has a flamboyant and colourful personality. He is usually very dignified in his manner and conducts himself with an air of confidence and authority. The Rooster is usually very ambitious, but can be unrealistic in some of the things that he hopes to achieve.

MIKE'S SURVIVAL — WALKING

Mike also needed his cerebellum for long-term survival. Your cerebellum helps you walk. If you are bed-ridden, you can get "bed-sores" or, in Mike's case, "straw-sores", which can get infected and lead to amputations. Being able to walk is almost essential for any extended life span.

The cerebellum coordinates the activity of your various muscle groups, to make your movements smooth. Strong muscles in your legs are only part of walking — you would immediately fall over if the cerebellum didn't "smooth out" and coordinate your various leg muscles.

Mike's cerebellum meant that he could walk without falling over — as long as his cerebellum was supplied with incoming information about "balance".

His single remaining ear provided that information. The ear does more than turn sounds into electricity, which it sends to the brain. It also tells you which way is up. The balancing mechanisms in Mike's ear, combined with an intact cerebellum, gave Mike the ability to walk.

MIKE'S NECK PECK

Mike still had a spinal cord. So the reflexes that were located in his spinal cord still worked. He kept the "neck peck" reflex for a while. In his early days of headlessness, he would still try to peck at the ground with his headless neck, even though he didn't have a beak or head. But he soon learnt not to hurt his exposed neck.

THE HIGH LIFE

Despite the fact that he didn't have a head, Mike had a good life. He slept comfortably in an old apple box with fresh straw.

In fact, Mike thrived without his head. Over his 18 months of headlessness, his weight climbed from about 1.5 kg to about 3.5 kg. He didn't show any signs of discomfort or pain. Mike couldn't see, but he could hear. He couldn't crow like a normal rooster because he didn't have a beak, but he could gurgle.

Various humane societies, including the American Humane Society, came to look at Mike. They decided that he was having a comfortable life, and that he

OTHER HEADLESS CREATURES

The Japanese eat huge quantities of raw fish. Some restaurants specialise in how quickly they can prepare the fish. On one occasion, a colleague of mine had a fish that had been "killed", cleaned and was ready for eating — the problem was that it was still moving.

And I have heard stories describing an axe head being dropped onto a cockroach, severing the body from the head. Over a few minutes, the body then walked several metres away from the head.

wasn't in pain, and that it would be cruel to kill him. (Anyhow, how would they kill him — cut off his head?)

However, Mike never "matured". His intellectual development came to an abrupt halt when his head was cut off. So even though he grew bigger, he still behaved like a crazy, mixed-up, "teenage", four-and-a-half-month-old chicken.

ON THE ROAD

Mike began earning his keep. First, the Olsens took him to California, and exhibited him on the boardwalk in Long Beach — near the snake lady (who was really a man), the two-headed cow, the legless man who did tricks on a motorbike, and the man from Borneo who ate live chickens. That first trip was a great success.

So Mike hit the road for a month or two at a time. It cost the public 25 cents to look at him. In one month alone, the gross earnings from Mike were US$4500 (about US$45 000 today). That works out to around 600 people per day paying to see Mike.

The Olsens used this money to lift their farm from the Horse-and-Mule Age to embrace the latest technology such as a tractor, a hay baler and a 1946–1947 Chevy pickup truck. (In 2001, Lloyd's grandson was restoring that same pickup.)

THE TRAGIC END

But it all came to an end in March, 1947.

Lloyd had stopped in a motel, on his way back from showing Mike in Phoenix, Arizona. Chickens, like all animals, generate mucus in their airpipe (trachea). A regular chicken will just lower its head, cough up the mucus and then swallow it down the oesophagus. But Mike couldn't do this — there was an air gap between his trachea and his oesophagus.

Whenever Mike choked on his mucus, Lloyd sucked it out with a rubber bulb syringe. Unfortunately, on this critical occasion, Lloyd had left the rubber bulb syringe back at the show in Phoenix.

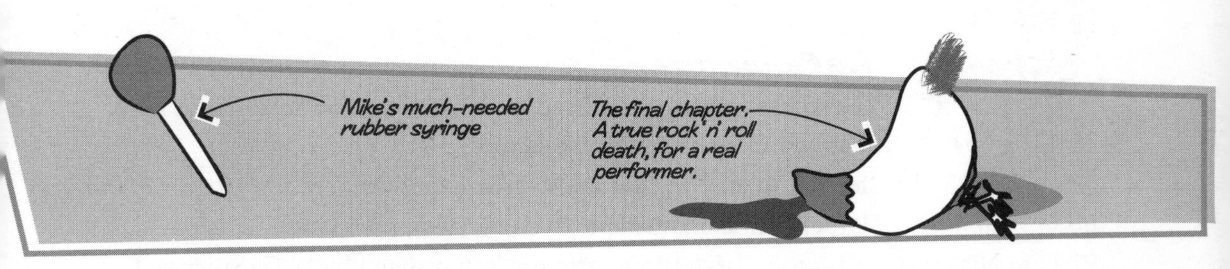

Mike survived without a head for some 18 months. He needed help, though. He was fed through an eye-dropper and he'd regulary choke on mucus or juices in his oesophagus, so a rubber bulb syringe was slid down his throat to suck the mucus out. All charming stuff. Mike's undoing came about when he drowned in own mucus after completing a "gig". His "support crew" had left the sucking syringe at the show ... they couldn't clear his throat ... and he met his maker!

He wasn't prepared to perform (or else he didn't know) mouth-to-mouth resuscitation (or should that be mouth-to-neck?), and so Mike finally died.

Miracle Mike died a true Rock Star death — choking on his own body fluids, in a motel room, on his way home from a gig in Phoenix, Arizona.

There was a chicken that had no haid,

How was it that this chicken weren't daid?

They cut his haid off way up high,

'Cause they wanted him for chicken pie.

Then poor ol' Mike had just a neck,

But lost his beak with which to peck . . .

So he had no haid,

But he could be faid,

And that is how he stayed not daid . . .

References

"Headless rooster — Beheaded chicken lives normally after freak decapitation by ax", *Life*, 22 October 1945, pp 53–54.

Dr Stephen Juan, *The Odd Body: Weird and Wonderful Mysteries of our Bodies Explained*, HarperCollins Publishers, Sydney, 1995, pp 53–56.

Richard Nickel et al, *Anatomy of the Domestic Birds*, translated by W.G. Siller and P.A.L. Wight, Verlag Paul Parey, Berlin, 1977, pp 20–22, 28–29, 34–35, 45–49, 114–124.

Teri Thomas, *The Official Mike the Headless Chicken Book — A 1940s Tale of Two Men and a Chicken, Fruita Times*, Colorado, 2000.

www.miketheheadlesschickenbook.com.

www.miketheheadlesschicken.org.

EGGSTASY

Q *Why are eggs egg-shaped? Why do they float when they're bad or old? How will spinning them tell you the difference between boiled and unboiled eggs? How come the sliced eggs at a buffet all seem identical? How do you get peeled eggs into — and out of — bottles? Is it possible to balance an egg on its end only on the Equinox?*

Before I could answer these questions (and others), I had to find out a lot of background information about eggs.

We humans eat hundreds of billions of chicken eggs each year. In the course of handling eggs over thousands of years, we have noticed odd things about them.

EGGS — A HISTORY

An egg is a self-contained environment that beautifully protects the life within during its earliest moments. Eggs let life come onto the land, hundreds of millions of years ago.

The eggs of fish could be simply squirted into the surrounding water. Amphibians laid eggs that could survive out of the water, but which still needed a moist environment. Reptile eggs were more resilient, but they too carried very little water. So reptile eggs still needed a damp environment.

It took another 150 million years for bird life to evolve. The bird egg can survive in a relatively harsh environment.

EGGS — AN INTRODUCTION

An egg is a chook's way of making another chook. She takes about a day to produce and lay one, and half an hour later she's ready to start to work on the next egg.

An egg has two main parts: a tiny amount of DNA which can grow into a living creature, and a whole lot of food to nourish that creature until it leaves the egg. All eggs have some kind of protective coating to help them survive — a hard shell, a leathery sac, a membrane, or just a mass of jelly.

Oviparous animals, such as chickens, dump their eggs to hatch *outside* their body. Bird eggs are among the largest eggs of any animal, because they have to carry all the food needed to grow the embryo into a chicken. The only extra ingredient needed to make a baby chicken is oxygen, which diffuses through the 7000–17 000 holes (or pores) in the shell.

Viviparous animals, such as mammals, have eggs with soft coatings that grow and mature *within* the mother's body. These animals eventually give birth to *living* young. The eggs don't have to be as large as bird or reptile eggs, because the growing embryo gets its food and raw materials from its mother. However, one order of mammals, the monotremes — the platypus and echidna — deposit their eggs outside the body.

WHY EGGS ARE EGG-SHAPED 1 — ROLL BACK?

First, what is the shape of a chicken egg? It's not spherical, like some reptile eggs. It's not a three-dimensional oval either. It's an asymmetrical mix of oval and tapered, with one end bigger than the other.

If eggs were rectangular little boxes, they would be very strong on the corners, but very weak in the middle of the straight walls. They would also be extremely uncomfortable for the chicken to lay.

The strongest shape of all is a ball, or sphere. But if a spherical egg were nudged, it could roll away downhill, never to be seen again.

So one reason why eggs have an "asymmetric tapered oval" shape is that if you nudge them, they'll come back to you. They'll sweep out a circle around the pointed end, and come to a stop with the pointed end facing uphill. In fact, the eggs of birds that have their nests on cliffs are more oval than the eggs of birds that nest on the ground. So the "more oval" eggs of these cliff-nesting birds will roll in a very tight little circle, and be even less likely to roll out of the nest and off the cliff.

Another reason for eggs to be egg-shaped is that they fit together quite snugly in the nest, with only small air spaces between them. This means the eggs radiate their heat onto each other, and keep each other warm. And of course, you can fit more eggs into the nest.

But the real reason why eggs are egg-shaped is so they can fit exactly into egg cartons and those cute little egg holders in the fridge!

WHY EGGS ARE EGG-SHAPED 2 — SQUIRT OUT?

Another reason why eggs are tapered is so that they can get pushed out of the hen. It sounds intuitively "wrong" and extremely uncomfortable, but eggs are laid with the blunt end coming out first, followed by the tapered end. In fact, the physics of pushing-an-egg-out tells us that eggs have to come out blunt end first.

Think how easy it is to squeeze a tapered cherry pip between your fingers and squirt it a metre or two. Think how hard it is to squirt a cube-shaped dice anywhere if you just squeeze it between your fingers.

The end of the chicken egg that tapers to a point has the ideal shape for the muscles of the hen's vagina to push on. As the muscles push towards the centre of the vagina, they have a perpendicular component of force towards the outside world — thanks to the wedge effect. But imagine that the vagina muscles were trying to push on something with flattened sides like dice (or the blunt end of the egg). No matter how hard the muscles push towards the centre line of the vagina, very little of that force would push the dice towards the outside world.

I wonder if this has anything to do with the shape of human babies, who are usually born head-first? They're biggest at the head and shoulders, and are thinnest at the legs.

CHICKEN MAKES EGG

When female chickens are tiny embryos in the egg, they start growing two ovaries. But by the seventh day of incubation, the right ovary fails to develop. However, some birds, like the kiwi and many birds of prey, have two functioning ovaries.

In the mature chicken, only the left ovary still works.

It is difficult to get a chicken to generate one egg nearly every day. You have to give them the right ratio between light and dark (about 17 hours of light and 7 hours of dark), and a diet that is high in protein and fat. Under this regimen, they will come in to "lay" at 22 weeks of age. They will then generate almost one egg per day, until they are killed one year later at 72 weeks (this is as long as they are able to keep producing this number of eggs). The chickens walk a "physiological tightrope", having to generate such a massive output of eggs. They are now "machines" that turn food into eggs. So laying chickens have huge amounts of fat in the abdominal cavity, as well as fatty infiltration of the liver, which is virtually a state of disease.

WHY EGGS ARE EGG-SHAPED 3 — SUPPOSITORIES?

The principle that hens use to lay eggs is also used in suppositories.

Suppositories are little tapered, egg-shaped drug capsules designed to be inserted into the anus. They are given PR (PR = per rectum = via the rectum) when there are problems with other drug-delivery methods. For example, a small baby with a dangerously high fever will sometimes spit out the paracetamol you put in their mouth to bring down their temperature. So medical doctors will deliver the paracetamol PR, via a suppository. The drug is absorbed directly through the wall of the anus into the blood vessels, and then into the general circulation.

Just as an aside, practically everywhere along the 10-metre length of the gut, the blood vessels go first to the liver before they go into the general circulation. This is so that the liver can remove "added" chemicals in the blood — usually food chemicals, such as fats, proteins and carbohydrates. In low concentrations, these food chemicals are essential for life. But if these chemicals reach high concentrations in the general bloodstream they are toxic. In certain liver diseases, where the liver does not remove these chemicals, the brain consequently suffers.

But the plumbing is quite different for the blood vessels at each end of the gut — at the anus, and under the tongue. Medical people see the gut as reaching from your mouth to your anus. These blood vessels don't go to the liver — they feed directly into the general circulation.

Paracetamol suppositories work PR, because the drug can get quickly into the bloodstream without being changed. Paracetamol also works if the patient swallows it. The paracetamol survives the trip through the stomach acids and the liver. There would be a time delay before it started working, but it would still work.

On the other hand, if you were to swallow nitroglycerine, it would not survive. It would get broken down before it reached the general circulation. (Surprisingly, nitroglycerine-the-explosive is also a medicine. It relieves the angina pain of heart disease, when you place a small dose under the tongue.)

If you gently push in a paracetamol suppository tapered end first, the baby will immediately (and frustratingly) push it out. But if you insert it blunt end first it stays in and begins to dissolve, reducing the baby's fever and/or pain.

OVARY TO LAID EGG

In animals, the ovary is an organ that continually makes follicles. When a follicle is expelled from the ovary, it is then called an ovum. It travels through the oviduct to the cloaca, and is then laid. But the oviduct is much more than a simple pipe joining the ovary to the cloaca — it is a manufacturing and packaging plant.

There are five regions in the oviduct.

The first region is the funnel of the oviduct, the infundibulum. This muscular section "catches" the egg as it leaves the ovary. In laying hens, it's about 7.4 cm long. The egg spends about 20 minutes here. The infundibulum feels a little like plastic cling film. This is where the rooster's sperm live, and where fertilisation may occur. By the way, rooster sperm can survive for 30 days in the infundibulum. The longest that human sperm can manage is only three or four days.

The second region is the "regular" oviduct, the magnum. It's the longest section, at 32.5 cm long, and holds the egg for about three hours. The magnum is a glandular region. Here, two layers of albumen are laid down around the central yolk. About 40–50% of all the albumen gets added in the magnum. The remaining 50–60% is supplied by the next two sections, the isthmus and the uterus.

The third region is the narrow part of the oviduct that leads into the uterus, the isthmus. The egg takes 70 minutes to get through this short section, only 8.7 cm

Before crossing roads, we need an egg!

We humans eat many, many eggs ... and in a huge variety of ways. It is estimated that we consume hundreds of billions of eggs a year.

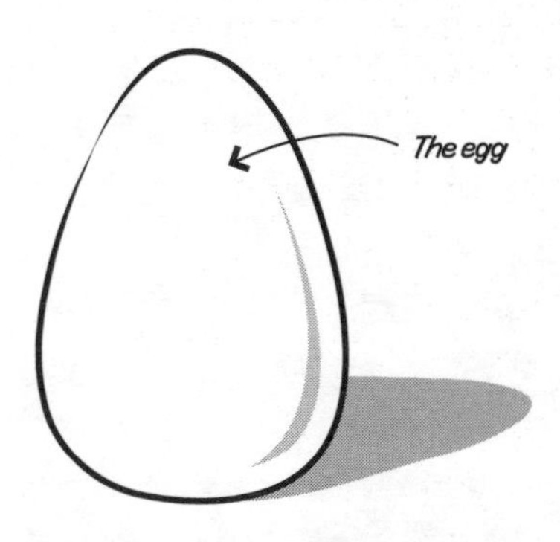

An egg is a self-contained environment that protects the life within during its earliest moments.

long. Here, water is added to the albumen, and the inner and outer shell membranes are deposited around the yolk. These membranes are made from a complicated mix of fibrous protein. The air space that exists in eggs is between these inner and outer membranes. As the egg gets older, it loses water through the pores — so the air space gets bigger.

The uterus, or "shell gland", is the fourth region of the oviduct. Even though this section is only 8.3 cm long, the egg spends about 19 hours here. The hard calcium shell is added here. The hard shell provides strength, but is also porous enough to allow just the right amount of water loss and gas exchange.

The last region of the oviduct is the vagina. It's the shortest of all, only 7 cm long.

The egg finally leaves the hen via the "cloaca". This is a Latin word meaning "sewer". While female humans have separate external orifices for urine, faeces and copulation, female chooks have a single all-purpose orifice to the outside, the cloaca.

However, the laying of the egg is very sophisticated. As the uterus pushes the egg outside the body, it virtually turns itself inside out — and then contracts back in. As a result, the egg does not touch any faeces on the way out.

In terms of pain, the laying of the egg is more like a bowel motion than a human giving birth. After a chook lays an egg, she stands quietly for a few moments, rolls the egg over and then continues with the pleasantries of chook life.

COLOUR OF SHELL

The colour of the shell depends entirely on the breed of the chicken.

White eggs come from hens with white feathers and white ear lobes, such as the Single-Comb White Leghorn.

Brown eggs are laid by hens with red feathers and red ear lobes, such as the Rhode Island Red, Plymouth Rock and New Hampshire. These hens are usually slightly larger, and need more food — so brown eggs are usually more expensive. Brown eggs are more attractive to some fans of health foods, which is another reason they can carry a premium price — although there's actually no difference in the nutritional value.

The brown colour comes from a protoporphyrin called ooporphyrin. This chemical comes from the natural breakdown of haemoglobin, which is the chemical that carries oxygen in the blood.

The Araucana Hen is very special — she can lay eggs that are either blue, green or pink. The colour pigment that causes the blue or green colours also comes from the breakdown of haemoglobin.

WHAT HAPPENS WHEN YOU BEAT EGG WHITES?

When you beat egg whites, the volume can increase eight times. The liquid is replaced by bubbles, which are basically liquid surrounding an air space. These bubbles are essential for a meringue or soufflé.

The bubbles in beaten egg whites happen because the beating action stretches and uncoils proteins strings. Proteins are made of amino acids strung together. The "string" is not straight, but all coiled up because of electrical charges on the individual amino acids.

Another thing to realise is that small bubbles are stronger than big bubbles. Think how hard it is to blow up a balloon through the early "small" stage. But once the balloon gets bigger, it becomes easier to blow it even bigger.

The beaten egg white can have three useful stages. You should stop beating at the third stage, and not enter the fourth stage.

The first stage of beating egg whites is the Foamy Stage. This happens after a short period of beating. The bubbles are large and weak. The eggs whites are a mix of fluid and froth. If you pull out the beater the egg white follows it up — and then collapses. This egg white will not "hold a peak".

The second stage is the Soft Peak Stage. As you have beaten further, you have divided the big bubbles into smaller bubbles. The foam is white, shiny and flowing. As you lift the beater up out of the foam the egg white now forms a peak for a brief moment — and then the tip of that peak folds over. The proteins in the liquid walls of the bubbles have not yet stretched to their maximum. So this foam can still expand if you cook it in the oven.

The final useful stage is the Stiff Peak Stage. The foam has reached its maximum size. The proteins cannot be stretched further. If you slice the foam with a knife, it creates a narrow groove with stiff walls. If you lift the beater the egg white follows — and when it stretches and breaks loose from the beater, the tiny peak that is formed is stable. This is when you definitely should stop beating.

The fourth stage is the Dry Foam Stage. The proteins are now breaking up. The foam is no longer moist and shiny, but dry and clumped up. The clumps are unstable and prone to collapsing. If you lift up a beater the foam again collapses early in the lifting process, as in the Foamy Stage.

WHY BEAT EGG WHITES IN A COPPER BOWL?

If you want the egg whites in your pavlova moist, stiff and slightly golden, you should use a bit of copper. The French writer Denis Diderot was the first to write about this back in 1771.

There are many explanations for why this is the case. One is that the foam is stabilised by disulphide bridges. Copper stimulates the production of these bridges.

Another explanation was discovered in 1984. This work was done at Stanford University by the McGee family. Dr Mrs McGee has a PhD in biochemistry, while Dr Mr McGee has a PhD in literature.

One of the 40 or so known proteins in egg whites is conalbumin. Conalbumin is a delicate molecule. When it gets fully stretched across the skin of the foam bubbles in freshly beaten egg white, the conalbumin molecule tends to fall apart. But tiny amounts of copper can combine with conalbumin to make it much stronger — so it doesn't fall apart. The combination of copper-and-conalbumin is slightly yellow, giving properly whipped egg whites that classic golden tinge.

Another advantage of having a stronger bubble wall is that the bubble helps to hold the water in. This makes the foam more resistant to over-beating. As good chefs know, this can happen all too easily when you've already whipped up a moist, stiff egg white. If you beat it for another 60 seconds, it goes all floppy on you.

The McGee family found that a touch of copper helps. Egg whites whipped in glass bowls were grainy and dry after only 10 minutes. But egg whites whipped in a copper bowl were still moist and stiff after 20 minutes. So everybody should use a bit of copper to get nice, stiffly beaten egg whites.

If you are unemployed, you can use a worn fork that in better days was plated with brass and silver (you know, the ones you get from your local "op shop" for 5 cents each). This works because there is a touch of copper in the brass plating.

If you have a part-time job, you can afford a copper bowl or a copper whisk.

And if you have a full-time job, you can supply your chef with a sterling silver bowl. This works because sterling silver is 5% copper.

EGG MAKES CHICKEN

If the creature is hatched to live by itself when it is still small (like a sea urchin), then its yolk is quite small. But if the creature does most of its growth in the egg (like most birds), then its yolk is much larger.

The chicken eggs that we eat have not been fertilised by a sperm. But if the egg had been fertilised by a sperm, it would turn into a chicken.

The chicken yolk is so big because it has two jobs related to making chickens. First, it is a source of the chemical building blocks, which are reassembled into all the organs that make up a newborn chicken. Second, it is also the source of all the energy for the process of turning a single cell into a baby chicken with quadrillions of cells.

It takes a lot to make a yolk. The yolk is a complicated mixture of proteins, fats and carbohydrates. These foods are made in the liver and then carried in the bloodstream to the ovary, where they are assembled to make the egg. One of the hormones controlling this process, oestradiol–17 beta, is made in the ovaries.

The colour of the yolk comes from pigments in the diet of the hen, such as lutein, carotin and ovoflavin.

If the egg has been fertilised, the yolk is later broken down by enzymes in the membrane around the yolk. The chemicals from this breakdown process are carried in blood vessels to the growing embryo. Even some of the eggshell gets reabsorbed, to give the embryo some calcium salts.

STORING YOUR EGG

In general, chemical processes happen faster at higher temperatures. An egg will age more in one day at room temperature than in a week in the refrigerator.

But you should keep them in their carton. Eggshells are very porous with their 7000 to 17 000 tiny holes. The carton traps a layer of still air right against the shell. This slows down the rate at which gases and water move through these holes.

Another advantage of the carton is that it helps stop the egg from absorbing odours from elsewhere in the fridge. Eggs that smell like leftover cat food or decaying vegetables are not attractive.

If you keep eggs in their cartons in the fridge, and not in those little egg holders, you can safely store them for four to five weeks after their packing date.

What should you do with dirty eggs? If you wash them, you remove the protective surface coating, so that bacteria can more easily invade the egg. So don't store them — wash them and use them immediately.

WHY CAN YOU BALANCE AN EGG ON ITS END ONLY AT THE SPRING EQUINOX?

This claim is complete rubbish. It's equally easy, or difficult, to balance an egg on its end at any time of the year. Some TV reporters, who love a feel-good story for the end of the news, do the Spring Equinox egg balance "phenomenon" every year. They offer nonsensical explanations such as: *"The gravity of the Earth, Sun and Moon all line up, helping the egg balance."*

Equinox means "equal night". So the Equinox is a day, normally around 21 March or 21 September, where there are 12 hours each of light and of dark. This exact date varies a little bit from year to year, and also according to your latitude on the planet. And of course, the Earth's atmosphere bends the light of the Sun and tricks you. When the Sun is just "visible" above the horizon, the disc of the Sun is in fact fully below the horizon. So you can "see" the rising Sun for several minutes before it actually crosses the horizon. But ignoring all these confusing factors, the day of the Equinox is roughly halfway between the shortest and longest days of the year.

A typical Equinox egg claim would be worded: *"During the Spring (Vernal) Equinox (about 21 March), it is said that an egg will stand on its small end. Although some people have reported success, it is not known whether such results were due to the Equinox or to the peculiarities of that particular egg."*

This is truly a modern myth. Annalee Jacoby, the China correspondent for *Life* magazine, started this myth with an article in the 19 March 1945 issue of *Life*. She had seen and photographed a curious local Chinese ritual in the city of Chunking. The locals would balance eggs on one end only on the first day of spring, to which they gave the special name, Li Chun. They believed that the first day of spring happened about six weeks before the Spring Equinox, in early February. (Note that this date is quite different from most Western countries, which claim that spring begins on the Equinox). The local legend was that it was much easier to balance an egg on end on Li Chun than on any other day of the year.

Albert Einstein, one of the greatest physicists of the 20th century, was asked for his opinion on this ability of the egg to stand on end only on the Equinox. He said that it was rubbish. Then the 19 March 1945 issue of *Life* ran a large photo spread showing many eggs balanced on end. This "proof" that the Great Man himself, Albert Einstein, was wrong made fantastic publicity.

The story was picked up by United Press and sent out on the wire across the world. And the myth that you could balance eggs on end only at the Equinox went right along with it.

Mind you, the six-week gap between the Chinese start of spring and the Western start of spring was conveniently ignored. So the roots of this legend are flawed right from its beginning.

Even so, the public embraced this myth. For example, on 20 March 1983, Donna Henes, an "artist and ritual-maker", organised 100 people in New York City into standing their eggs on end in public. The *New Yorker* magazine published their coverage of this event in their April 4 issue. In 1984, 5000 people turned up at the World Trade Centre, desperate to stand their eggs on end to celebrate the Spring Equinox.

The myth spread further with each passing year. On 19 March 1988, *The New York Times* ran an editorial with the headline "It's Spring, go balance an egg". On 21 March, they ran a picture of people at the World Trade Centre standing their eggs on end.

Phil Plait has well and truly debunked this myth, via his *Bad Astronomy* home page. He works at the Physics and Astronomy Department at Sonoma State University, part of the California State University system. His current project is a NASA-sponsored public outreach program for a satellite named GLAST (Gamma Ray Large Area Space Telescope).

He managed to balance a whole bunch of eggs on end on 25 October 1998 — which is definitely not the Equinox.

Many others have joined him. Lisa Vincent, a teacher at the Mancelona Middle School in Michigan, gave her students the same task.

On 16 October 1999, she and her students were all able to balance their eggs — again, not on the Equinox. They wisely set up their experiment on sturdy school chemistry workbenches. Very impressively, they were able to balance their eggs on the more difficult pointy end, rather than on the easy rounded end. Some of their eggs remained standing for over a month, until random vibrations knocked them over.

A middle school in Tucson, Arizona, also tried it. They made two findings. First, they could balance their eggs at any time. Second, middle school students could successfully handle an egg about 36 times before they cracked it.

In fact, Phil's email mailbox is overflowing with stories and pictures showing eggs balancing at any time of the year. So if you can balance an egg on end, astronomy has nothing to do with it.

In balancing an egg, a little inside knowledge helps.

Part of the difficulty is that the yolk is usually a little off-centre. So you can help things along (= cheat a little) by shaking the egg fairly vigorously. This breaks the yolk loose from the bands of chalazae that locate it roughly along the centre line of the egg. This does two things. First, the yolk drops a little, lowering the egg's centre of gravity and making your job easier. Second, as the yolk drops, it gets closer to the centre line of the egg, again making the egg easier to balance.

There's one more factor in our favour. Eggshells are not perfectly smooth. They have microscopic bumps, which can act as tiny legs to hold the egg up.

So even though at the Equinox the hours of light and dark might be fairly closely "balanced", this has nothing to do with "balancing" an egg.

It all goes to show the wisdom of the Nobel Prize winner Richard Feynman, when he said, *"Science is a way for us to not fool ourselves."*

ANATOMY OF THE CHICKEN EGG

In your typical 58-gram egg, the shell weighs 6 grams, the white 33 grams and the yellow yolk 19 grams.

The yolk carries the microscopic DNA of the mother. This yolk is actually made up of alternating layers of white and yellow yolk. The vitelline membrane wraps around the yolk. (The only covering that worm eggs have is this vitelline membrane.) The yolk is about 50% water, 17% protein, 1% carbohydrate and 1% minerals, with the remaining 31% being fat.

The yolk is held in place by two "chalazae" (pronounced ka-LAY-zee). Each chalaza is a spiral strip of concentrated egg white tissue, running from the yolk to the internal membranes at each end of the egg. The chalazae keep the yolk roughly centralised, while leaving it free to rotate. They first develop in the isthmus. The yolk rotates as it moves along the oviduct, so the chalazae become twisted into a spiral.

You sometimes find these chalazae as stringy, rope-like strands in your cooked egg. They are more prominent in fresh eggs. It's perfectly fine to eat them. However, if you want your custard to be extra smooth, remove them by straining the egg mixture through a sieve.

There is a very slight depression on one side of the yolk, the "germ spot". This is where the sperm enters the yolk to fertilise it. This side of the yolk is less dense than the rest of the yolk. Thanks to the chalazae, the yolk rotates freely to let the germ spot swing around to the top — nearer to the warmth of the roosting mother hen.

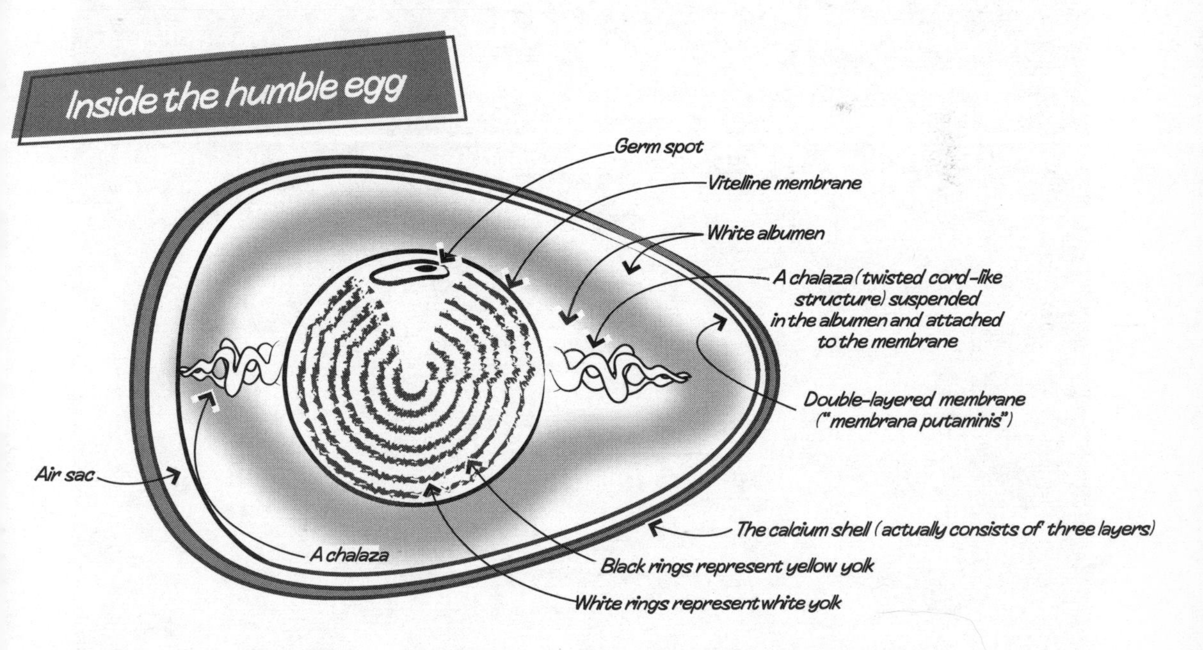

n egg is a chook's way of making another chook. The hen takes about two days to make one, and half an hour later she's ready to make another. An egg has two main parts: a tiny amount of DNA (which can grow into a new creature) and a whole lot of food to nourish the creature until it leaves the egg. Oh ... it also needs a protective coating (shell, membrane, sac or mass of jelly).

Around the yolk there is a small volume of solid albumen, which in turn is surrounded by a bigger volume of liquid albumen. The solid and liquid albumen together make up the egg white. The egg white is an extra source of food for the developing chicken. The white is about 88% water and 11% protein, with the remaining 1% being carbohydrate and minerals.

A tough double membrane, the "membrana putaminis", wraps around the egg white. This double membrane is made up of an inner and outer shell membrane, as described earlier. When the egg is laid, it is warm and moist. It immediately begins to cool. Air can cross the outer membrane, but not the inner membrane — and so an air sac forms.

Finally, the membranes are covered by the hard calcium carbonate shell, which is made up of three separate layers. Even so, the shell is only about 0.32 mm thick. It's about 95% minerals, 3% protein and 2% water.

"THEY COULDN'T EVEN BOIL AN EGG"

1 Actually, cooking an egg in its shell is quite tricky. The first surprising thing to realise is that you should not have the water too hot.

The egg white will turn solid between 60 and 65°C. The yolk has different proteins, causing it to solidify between 65 and 70°C. So you can make an egg go solid at temperatures a lot lower than 100°C.

If you leave an egg in boiling water for a few minutes, the delicate proteins get tangled and the texture becomes "rubbery". So cover the egg with warm water, bring it to the boil and then immediately turn the heat off. Then cover the pot, and let it stand for four to five minutes if the egg was originally at room temperature, or six to seven minutes if you took the egg from the fridge.

Not only will the eggs have a nicer texture, they'll also be easier to digest.

2 If you want to cook an egg in its shell (don't even say "boil an egg"), try putting a tiny pinhole at the big end.

As the egg heats up, it expands and pushes the air out. The air usually escapes through the tiny pores in the eggshell — but sometimes it doesn't. Having a pinhole guarantees that the egg won't burst open and turn into stringy white and yellow strands.

There's another advantage to this. The air space will collapse as the liquid egg white expands. So you'll end up with an oval egg that doesn't have a dent at one end — a fairly obscure aesthetic advantage, mind you ...

CAN A DOUBLE-YOLKED EGG PRODUCE TWO CHICKENS?

Folklore says that an egg with no yolk at all is unlucky, because it was laid by a cock. A double-yolked egg is also supposed to be unlucky, because it implies an imminent death in your family. But a fertilised double-yolked egg . . . ?

There are three main options as to what can happen when a follicle is generated by the functioning left ovary of a chicken.

In the first option, everything goes normally and one day later the chicken lays that egg, complete with shell.

In the second option, the yolk does not get picked up by the infundibulum (the funnel of the oviduct). Instead, it ends up in the abdominal cavity, and is eventually reabsorbed. This happens from time to time in cockatiels. But as part of the process of the reabsorption of the egg, massive amounts of fat appear in the blood. As much as 30% of the volume of the blood will be fat! This causes a process called "sludging", where the blood suddenly contains large blobs of fat. These blobs can block the flow of blood to important parts of the brain.

It's fairly common for the affected cockatiel to suffer a mild stroke. The characteristic sign is that she lays her head over to one side. But magically, she somehow recovers within a day from her mini-stroke — something that we humans sadly cannot do.

The third option is that the ovary will generate two yolks at the same time. Unfortunately, these eggs rarely hatch.

In a hatched-and-sat-on-by-Mum egg, the unhatched chicken has to rotate itself so it get its head up to where the air cell is, between the inner and outer membranes at the larger end of the egg. Then, under normal circumstances, the baby chicken will peck its way out.

But if there are two chickens inside, they will almost invariably fight each other. Neither of them will be able to get to the air cell, so they will both die.

However, there have been a few very rare cases where the egg has been carefully opened at exactly the right time, and two chickens have survived from a double-yolked egg.

COLOUR OF YOLK

The colour of the yolk of an egg depends on what the hen eats.

The egg companies are not allowed to add "artificial" colour additives. But they *are* allowed to add dried marigold petals (usually from Argentina). These contain natural intense yellow-orange pigments (called xanthophylls), which enhance the yellow colour of the yolk.

If the hens are fed yellow corn and alfalfa meal, their yolks will be a medium yellow. Barley and wheat will give an even lighter yellow. But if you feed your chickens a totally colourless feed, such as white corn, the yolks will be white.

WHY IS EGG WHITE SOMETIMES CLOUDY, OR GREEN-YELLOW?

An egg is a dynamic, breathing thing. In a very fresh egg, the carbon dioxide sometimes has not had enough time to escape. The trapped bubbles make it a little cloudy.

About 0.6% of the egg white is made of various minerals and chemicals. If there's a lot of riboflavin (vitamin B2), this can give a greenish or yellowish tinge to the egg white.

PHYSIOLOGY OF THE CHICKEN EGG

The inner and outer shell membranes, as well as the hard shell itself, are all permeable in varying degrees to gases and water.

Soon after the egg is laid, water leaves and air enters through the pores in the shell. During the process of incubation, bird eggs lose about 15% of their original weight as water evaporates through the shell. The air accumulates in the space between the two membranes at the larger end of the egg. So as the egg gets older, the air space gets bigger.

Bird eggs tend to lose the same amount of water each day. But they increase their consumption of oxygen and production of carbon dioxide as the embryo grows.

EGGS AS FOOD

If an egg gets fertilised, it can grow into an animal or insect — which you can potentially eat. But you can also eat the unfertilised egg of a bird.

Chicken eggs are the eggs we humans most commonly eat. But we also enjoy the eggs of other birds, such as ducks and turkeys. The Chinese are very fond of salted and preserved duck eggs. Other birds' eggs that make it to the table include those of plovers, pheasants, ostriches, geese, pigeons, quails, gulls and lapwings.

Caviar is the name we give to the eggs (or roe) of fishes such as sturgeon and salmon. This delicacy is very highly prized and expensive. We also eat reptile and insect eggs.

HISTORY OF EATING EGGS

People have been eating eggs for thousands of years. Four thousand years ago the ancestors of the modern hen, Indian jungle fowls, had spread throughout China, the Middle East and Europe. However, they were prized for their meat more than their eggs. Columbus took chickens with him to the New World on his second voyage in 1493. There are now about 200 different breeds of chickens known.

The poultry industry expanded very rapidly during World War II, because of the need to deliver a lot of meat and other proteins very quickly. Today, thanks to "improvements" in feeding and breeding, it takes only six weeks to produce a full-grown broiler and five months for a laying hen.

Your average modern poultry farm could have over a million laying hens. These hens are incredibly efficient egg factories. They need about 1.8 kg of feed and 3.4 kg of water to make one dozen eggs. Back in the 19th century, an egg-laying hen would have produced about 100 eggs per year. But today's typical egg-producing hen delivers about 280 eggs per year.

In 1990, China produced about 160 billion eggs, the European Union about 86 billion eggs and the USA about 68 billion eggs.

CAN YOU EAT RAW EGGS?

There are two main potential problems with eating raw eggs: biotin deficiency, and food poisoning.

One of the many proteins in the egg white is called "avidin". Avidin is broken down by the heat of cooking, but it survives in raw eggs. Avidin binds very well to a chemical in your body called "biotin" (also known as vitamin H).

If you eat between 3 and 12 raw egg whites (without the yolks) each day, you can suffer a biotin deficiency. This can give you dry skin, hair loss, loss of appetite and tiredness.

However, there is biotin in the yolk of the egg. So if you eat the complete raw egg, much of the avidin in the white will be mopped up by the biotin in the yolk — helping you avoid a biotin deficiency.

Food poisoning can happen because of bacteria (such as salmonellae) that live naturally in the gut of the hen. The eggs come out of the same hole as the faeces — but because the uterus turns itself inside out, there is usually no contact of egg with faeces. The eggs are slightly porous, so there is always the risk that, if the egg is laid onto faeces, salmonellae bacteria can enter the egg. If they did, they would be killed by cooking — but they could survive in a raw egg.

ALBUMEN vs ALBUMIN

"Albumen", with an "e", is the name given to the egg white.

"Albumin", with an "i", is a general name given to a class of simple proteins that dissolve in water and also coagulate when heated. There are many "albumins" — serum albumin in the blood, lactalbumin in milk, myo-albumin in muscle, legumelin in the seeds of leguminous plants, and ovalbumin and conalbumin in the egg white. Other proteins in egg white include ovoglobulin, ovomucin and ovomucoid.

The egg white also contains special proteins, lysozyme and ovotransferrin, that slow the growth of bacteria. They protect the yolk, and the chick embryo in a fertilised egg, from invasion by bacteria.

However, bacteria are many and these proteins are few — and they only slow, not stop, bacteria. So bacteria can grow in egg whites. If you have any egg whites left over from a recipe, cover them and store them in the fridge for a maximum of two days. Luckily, egg whites freeze very well. So freeze them in little zip-lock bags or in ice-cube trays, and then transfer them to plastic bags. That way, you defrost only as many as you need.

(In 1968 The Beatles cracked into the charts with *The White Albumen* and the chicks went wild.)

FOOD CONTENT OF EGGS

Your average 56-gram egg contains about 6 grams of protein (about 15% of your average adult's daily protein needs) and about 12 grams of fat. This protein contains all the essential amino acids that we humans need. Eggs also contain iron, and all the vitamins except vitamin C.

EGGS IN COOKING

Whole eggs can provide leavening (expanding) properties, texture and moisture when they're used in cake batters. You can also coat foods in breadcrumbs or flour by first dipping them in beaten eggs. Eggs help bind foods together.

Eggs can contract during cooking. This property is used in croquettes and terrines to preserve their shape.

Eggs can be boiled, baked in their shell, scrambled, poached or fried. They go well with a range of sweet or savoury foods.

You can easily make French toast by soaking white bread in liquefied whole egg, and then pan-frying it until it turns a lovely golden-brown.

HOW DO THEY GET THE EGG SLICES AT A BUFFET BAR TO ALL LOOK THE SAME?

When was the last time that you ate a 30-cm, 500-gram hen's egg? If you eat at buffet restaurants a lot, the answer would be quite recently.

The trouble with a real egg is that to get egg slices not only do you have to boil and peel it, but you also waste the end bits, which are mostly white with almost no yellow. And there are some slices that are mostly yellow with hardly any white. It's all very unbalanced.

Enter the Gourm-Egg — 30 cm long, 500 grams in weight, invented in 1968 by Purina Ralston. It's your genuine, machine-hatched egg cylinder. It's used by hospitals, restaurants, airlines and people who go in for quantity feeding. It provides at least 75 slices.

The Gourm-Egg is also called the Long Egg. Among commercial foods, it's classified as an "extruded product". You start off with liquid egg whites and yolks that are already separated. The yellow yolk centre is cooked and then extruded to be the diameter of a normal egg yolk. This is covered with a thin layer of egg white and then frozen for storage. A small amount of starch or food gum (usually 1% or less) is added to the albumen to stop it "toughening" in the freezing process. By the way, just so that it doesn't look too perfect, they place the yolk slightly off-centre. The Gourm-Egg slices look pretty authentic, but you can sometimes pick them by little wrinkles along the edges of the egg slice. The wrinkles come from the plastic wrapping tube.

The Gourm-Egg was not the first egg "analogue". That honour went to a "product" that looked like scrambled eggs when cooked but had no cholesterol. It was made from various egg proteins and dairy products, with added emulsifiers, stabilisers, flavours and colours. The ingredients were blended, heated to kill bacteria, packaged and finally refrigerated.

Another brand new egg product was introduced into the USA in 1983: hard-cooked eggs with a three-month shelf life. The egg is hard cooked and then coated with a lacquer to seal the porous shell. These eggs can be kept for three months in the fridge, or one month at room temperature. People sometimes colour them for use as Easter eggs they can later eat, or take them on camping trips. Some of us might think that the Gourm-Egg is a little eggsessive. But I'm sure it's better this way — the hens would flip if they had to lay 30-cm-long eggs.

Get Egg Into Bottle

Professor Julius Sumner Miller popularised science — especially physics. He introduced millions of people to the awe and wonder of science. One of his famous demonstrations was getting a peeled boiled egg into a milk bottle.

HERE'S WHAT YOU NEED:

- **A boiled egg that has been peeled of its eggshell.** Boil the egg gently for about 10 minutes. If the egg is not hard enough, the white of the egg can sometimes slide off as the egg slips into the bottle. In fact, it's handy to have a few spare eggs, so you might as well boil up a few. Once you've boiled them, cool them down quickly in cold water, so you can peel them without burning yourself.
- **A way to set paper alight.** A box of matches, or a cigarette lighter, or a candle. Don't burn *yourself*.
- **A piece of paper, about half the size of a bank note.** You need to drop the burning paper, slightly scrunched up, into the bottle. This means that the paper should burn easily, and should not be too light. Newspaper is fine. On the other hand, tissue paper often won't burn with a flame, but will just smoulder — and if it does burn the tissue paper is so light that it will sometimes fly up into the air.
- **A milk bottle.** This is the hardest part — trying, in our disposable society, to get hold of an old-fashioned milk bottle. Actually, it doesn't have to be a milk bottle, but the bottle does need a few features:

1. The mouth (or opening) should be 3–10 mm smaller than the egg.
2. The mouth of the bottle should also be smooth, with no jagged edges. This means that the egg can slip in easily, without tearing or shredding itself.
3. The bottle should have a volume of at least 500 cc. The bigger the volume, the better — up to a few litres. The extra volume gives more room for the hot gases.
4. The bottle should have a body that is bigger in diameter than the mouth. This gives you a nice popping sound as the egg goes in. So if you want the sound effects, a jar with a constant diameter is no good.

If you can't find a milk bottle, you could still try a chocolate sauce bottle, or a fruit juice bottle. If you have emu eggs, you can use a bottle with a bigger mouth, and if you have really tiny eggs, you could probably use a beer bottle.

Don't use a plastic bottle, because you need to drop burning paper into it. If the plastic bottle catches on fire, it will be really hard to put out, the blobs of molten plastic will stick to your skin and give you really horrible burns, and the burning plastic will give off poisonous gases. A glass bottle is better, providing it doesn't break.

- **A sink.** The advantage of doing it all at the sink is that if anything goes wrong, you just drop the burning bits into the stainless steel sink and turn on the water.

MEDICO-LEGAL WARNING STUFF

If you want to do this experiment, take care. Don't burn yourself, or anybody else, with the burning paper or the water that you boil the egg in. Don't break the glass bottle, but if you do, don't cut yourself.

DOING THE EGGSPERIMENT:

First, set up your peeled boiled egg, your bottle, your piece of paper and your matches or cigarette lighter at the sink.

Scrunch up the paper, light it and drop it into the bottle. Pick up the peeled boiled egg immediately and place it carefully in the mouth of the bottle. As it burns, the carbon in the paper combines with the one volume of oxygen in the air in the bottle to make one volume of the gas carbon dioxide.

Now, here's an important bit of chemistry. One volume of oxygen will give you one volume of carbon dioxide — so there should be no change in volume. But you have a flame, and so this gas is hot. It expands, taking up more volume. The egg is sitting in the mouth of the bottle. So you should see the egg bounce up and down gently as the hot gases try to get out of the bottle.

Eventually, the flame goes out, which means that you're not making any more carbon dioxide. The egg stops jiggling up and down, and just sits there in the mouth of the bottle.

Gradually, the gases inside the bottle cool down and begin to shrink. This means that you have a slightly lower pressure inside the bottle compared with the normal atmospheric pressure outside the bottle. The egg is sitting between the high pressure outside and the lower pressure inside. The egg is flexible, so the high pressure pushes it into the bottle — usually with a satisfying "popping" sound.

Get Egg Out of Bottle

So how do you get the egg OUT OF the bottle, in one piece, without breaking the bottle? It can be done. In fact, I know of two ways to do it.

I'll give you a hint. As Obi-Wan Kenobi would have said, "*Use the Force, Luke, use the Force that pushed the egg INTO the bottle.*"

The force that pushed the egg INTO the bottle came from the pressure difference between the inside and the outside of the bottle. When the pressure was greater on the OUTSIDE, it pushed the egg INTO the bottle.

So, if you want to push the egg OUT OF the bottle, you need to have a greater pressure INSIDE the bottle. There are a few different ways to do it.

The first way to get a greater pressure inside the bottle is easy. Tip the bottle upside down, so the mouth of the bottle is pointing downward. The Law of Gravity makes sure that the egg rests in the mouth of the bottle.

Next, put your mouth up to the mouth of the bottle and blow gently, but firmly, INTO the bottle. The air coming into the bottle from your mouth, under pressure, will push the egg out of the way. Once you have dumped as much air as you can into the bottle, the pressure will be the same on the inside of your *mouth* and the inside of the *bottle*. Remove your mouth from the mouth of the bottle, and the excess pressure in the bottle will push the egg out. Problem solved!

The second way to get a greater pressure inside the bottle involves burning more paper inside the bottle!

Normal air contains about 21% oxygen. The air in the bottle has much less than 21% oxygen, so it probably would not support a decent flame. But the air that comes out of your lungs still has a lot of oxygen left in it — about 16%. That's why mouth-to-mouth resuscitation can work, because there's still heaps of oxygen left in the air coming out of your lungs.

So hold the bottle with its mouth upwards, with the egg in the bottom of the bottle. Then blow some air into the bottle, and then blow some more, and do this about a dozen times. You will soon bring the oxygen level in the bottle up to 16%.

Again, scrunch up a piece of paper, set fire to it and drop it into the bottle. Quickly, turn the bottle upside down, so that the egg rolls into the mouth of the bottle and seals it off. As the paper burns hot gas will be generated, and it will push the egg out of the bottle. Once again, problem solved! (Or eggsactly the result you were after! — Sorry.)

EGGSPLODING EGG ON YOUR FACE — 1 AND 2

In the hands of the talented amateur, the simple egg can be a dangerous weapon. This tale is told by a young man, JY, who now has his PhD in Physics. His story, and his insightful explanation, serve as a warning to those who belittle the power of the hot egg ...

"I carefully filled the container with water up to a level which guaranteed that the egg would be completely submerged, and turned the microwave on high. I sat down, and after about two minutes heard a loud crack followed by a hissing boiling sound. Said egg had self-destructed in a big way. Every single wall of the microwave's interior was drenched in egg.

Undeterred, I prepared another egg for boiling, this time putting much less water in the container and using a far lower microwave power. After five minutes of this reduced power, I inspected the egg and decided that the operation was a success, in spite of some rather large cracks in the egg's shell. After at least 30 seconds I began pulling the shell off the egg. Most of the shell was gone, and it was at least a minute after heating, when the egg exploded violently in my face.

I had large amounts of egg stuck in my hair and (more seriously) in my eyes. Also, the outline of my head and shoulders were discernible as a Blast Shadow on the wall behind where I was sitting. It really, really flew — part of the yolk made it to the wall on the other side of the room.

So, in my attempt to boil one egg, I:

- *covered the microwave, the TV room and myself with boiling hot egg;*
- *wasted 2 eggs, and have spent 1.5 hours cleaning and recounting the affair.*

I did manage to consume a small amount of extremely over-boiled egg when, immediately after the second explosion, I flung my head back in pain and fright in such a way that a piece of egg that had stuck to my nose dropped happily into my mouth.

My guess is that the egg contained a super-heated pocket of air which, on contact with a fresh piece of liquid egg (through me rolling the egg round in my hand for the purpose of peeling it) boiled the egg and caused the explosion."

In *The Inquisitive Cook*, Anne Gardiner describes an occasion when she wanted to warm a cold egg for seven seconds in the microwave, but accidentally dialled up one minute and seven seconds: *"Realizing the mistake, she flung open the microwave's door just as the egg exploded. Even the kitchen ceiling had to be repainted."*

Spin an Egg – Is It Boiled or Fresh?

You can "see" a difference between boiled and unboiled eggs. How? It's a three-stage process. Spin the eggs, stop them by touching them precisely and deliberately, and then look carefully.

HERE'S WHAT YOU NEED:

- **A boiled egg.** Same as for getting an egg into a bottle. Let it cool down. Don't peel off the shell.
- **An unboiled egg.** Leave its shell on (otherwise it leaks everywhere).
- **A marking pen.** Mark the unboiled egg. Don't add anything (like a big lump of masking tape) that might change either the weight distribution or the spinning properties of the egg.
- **A flat and level surface.** If the surface is bumpy, it will interfere with the smooth rotation of the spinning eggs. If it's not level, the egg might spin off onto the floor. You need a clear spinning area roughly the size of a big dinner plate.

MEDICO-LEGAL WARNING STUFF

I strongly recommend that you do nothing that would need medical care, and that you don't break any eggs, because then you'd have to clean up the mess. Now that we have that stuff out of the way, let's go.

DOING THE EGGSPERIMENT:

First, take the boiled egg, and place it carefully in the middle of the cleared area on your flattish, level surface. It should be on its side. Don't try to put it on one of the ends.

Now, using your more skilful hand (the hand you write with), reach over to the egg with your thumb and index finger, and give it a bit of a spin — and then remove your hand. Let the egg spin about five times. After about five spins, reach over, and — *here's the tricky bit* — touch the egg gently again with your thumb and index finger, but only for a second, so that the egg stops spinning — and then immediately let go, and look carefully to see what's happening.

The boiled egg should be solid all the way through. So when you first apply rotary force to the eggshell, all of the egg (from the eggshell to the centre of the yolk) should immediately start rotating. And when you touch the boiled egg to stop it, everything on the inside of the egg should immediately stop. And when you immediately let go of the egg, then it should stay stopped.

Now, do exactly the same (spin, stop, let go and look) with the unboiled egg.

The unboiled egg is different. It's slushy on the inside. When you try to rotate it, the egg will spin, but the insides will lag behind a little. Energy will be wasted in the form of friction. For the same amount of applied force, the unboiled egg will spin a little slower than the boiled egg. (It will probably even wobble a little more than a boiled egg. Sometimes you can use this wobbling method to pick the unboiled egg, but it's not very reliable.)

When you touch the unboiled egg to stop it, the shell stops — but the soft, jelly-like insides continue to spin inside the stopped eggshell. Slowly, the friction force from the spinning insides starts the unboiled egg spinning again.

So there you have it — the boiled egg stops, but the unboiled egg keeps on going.

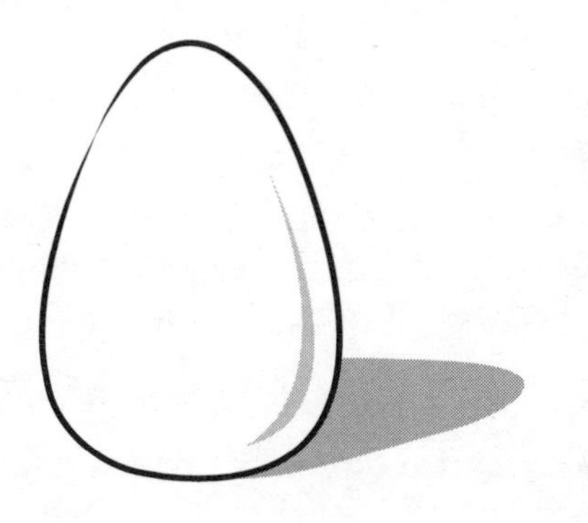

ALL EGGSHELLS, WEAK AND STRONG

The eggs of birds should be easy for the chick to get out of, but hard for predators to get into. According to Kenneth Entwhistle from the Manchester University of Science and Technology, that's exactly how they're built.

Imagine that you are pushing on one side of a thin and brittle material. In response, the other side stretches and distorts. If that other side is weak, a crack will form — and then spread back through the entire thickness of the material. It seems wrong, but the force that one side of a material can withstand depends on the strength of the other side.

Dr Entwhistle measured that the chicken eggshell is 34% stronger on the inside than the outside. So it's designed to stop invaders from getting in, but at the same time, to be easy for the tiny chicken to peck its way out of.

The inner surface of the chicken eggshell is made of strong calcite crystals arranged in a strong, honeycomb-like pattern. The outside layers, however, are broken up by organic materials, which makes them weaker.

EGG AND CHICKEN RECORDS

The largest chicken on record was a White Sully rooster called "Big Snow". He weighed 18.5 kg.

The longest flight by a chicken is thought to be 192 metres.

The record for laying the highest number of eggs is thought to be held by a White Leghorn, Chicken #2988, in an official test carried out by the University of Missouri. She laid 371 eggs in 364 days.

A White Leghorn in New Jersey laid the heaviest chicken egg on record. The double-yolked, double-shelled egg weighed 0.45 kg. You get a double shell when the egg gets stuck and spends longer in the uterus.

A Black Minorca in England laid the largest chicken egg on record. It had five yolks and weighed 0.34 kg. Its circumference was 31 cm around the long axis and 22.9 cm around the short axis.

The greatest number of yolks in a single egg was achieved in 1971 by a hen at Hainsworth Poultry Farm — nine yolks. Multiple yolks are possible when the ovary squirts out a bunch of smaller follicles that travel as a group down the oviduct, and in the uterus get encapsulated in a single large eggshell.

Howard Helmer, from the American Egg Board, is known as the "Omelette King". In 1990, he made 427 two-egg omelettes in just 30 minutes.

And the most expensive "egg" was the Fabergé "Winter Egg". It sold in 1994 for US$5.6 million. Carl Fabergé was the son of a Swiss immigrant to Russia. Carl was the Court Jeweller to the Romanovs, the Imperial ruling family of Russia, in the late 1800s and the early 1900s. He and his skilled jewellers made highly ornate mechanical eggs from precious materials such as gold and diamonds. The press of a concealed switch would set an elaborate clockwork mechanism (spinning globes, chirping birds, growing trees, etc) into operation. The "Fabergé Eggs" were given as presents between the various members of the Russian Royal Family — until they were executed by the Communists.

And of course we all know the chicken that survived the longest with his head cut off ...

"THEY COULDN'T EVEN FRY AN EGG"

There's no point in frying eggs in a sizzling-hot frying pan. It just overheats the proteins and causes too much "crosslinking" of the proteins, giving a tough, rubbery taste. The bottom of the fried egg will probably have a thin, tough, brown, crunchy membrane, while the white will be dimpled with many holes where the water was forced out.

The proteins all solidify by 70°C, so a gentle heat is all you need.

EGG WHITE AND COOKING

Egg white is mostly water, but the remainder is almost all protein with virtually no fat. Egg whites are only about 12% solid.

Albumen coagulates when it is heated. So egg whites are used in many foods to hold them together.

The protein of egg white has another useful property: it foams easily. But unfortunately, it is also quite weak. So it usually has to be supported during baking by another protein, such as the gluten in flour. Sponge cakes, angel food cakes, meringues and soufflés are leavened by beating the egg whites into a batter that contains flour (but definitely with no added raising agents) and heating them in an oven. The little bubbles expand. The proteins are now strong enough to maintain their structural integrity. A classic cake tin used to bake an angel food cake has a thick tube rising up from the centre. It's there to help the bubbles stay up.

You generally need a mixing utensil such as a whisk to break up the pockets of air into successively smaller bubbles. Smaller bubbles are stronger than bigger bubbles (see box, *What Happens When You Beat Egg Whites?*).

WHICH CAME FIRST, THE CHICKEN OR THE EGG?

Chickens make eggs, and eggs make chickens. So which came first?

In nature, evolution happens thanks to changes in the DNA of a fertilised egg. At one stage, there were birds, but there were no chickens. At some stage later, there were chickens. So in between, some non-chicken birds mated and produced an unusual egg. This egg led to the first chicken.

So if my logic is correct(!!), the egg came first ...

Of course, there's the famous cartoon that backs up my theory. It shows a chicken and an egg in bed together. The chicken looks quite unhappy, but the egg is smiling and smoking a cigarette ...

Unfortunately fats can collapse this foam — and there are small amounts of fats in flour. You can get around this by folding the flour into the egg white foam fairly gently. This minimises the contact between the fat and the protein.

CAN YOU EAT COOKED EGGS?

In 1991, the *New England Journal of Medicine* reported the case of an 88-year-old man who had eaten 25 eggs for every day of the previous 30 years. In the Lottery of Life, he had been given some very nice biochemistry. His cholesterol levels were low, and he had no signs of atherosclerosis (hardening of the arteries).

He had a *"great reduction in the efficiency of cholesterol absorption from the intestine, and ... a marked increase in the conversion of cholesterol into bile acids. These physiologic adaptations would leave little if any of the dietary cholesterol to elevate plasma cholesterol levels and be deposited in arterial walls."* In other words, he didn't absorb cholesterol as easily as most people — and he got rid of most of the cholesterol he did have by turning it into bile acids.

Egg whites are also used to make nougats. They are whipped into a mixture consisting of sugar syrup, fats and a milk ingredient. Here, the egg whites act as a foaming agent. (Sometimes, modified whey is used as the foaming agent.) The texture of the nougat depends on how much air has been whipped into the sweet mixture.

Foams made with egg whites tend to be more elastic, while those made with egg yolks tend to be more crumbly and tender. This is because the fat of the yolk tends to partially collapse the foam.

Egg white is also used in ice cream and sorbet. It acts as an interfering agent, and stops large ice crystals from forming.

SIZE OF EGGS

The size of eggs is worked out by weight, not volume.

Different breeds tend to lay different sizes. Leghorns tend to lay large eggs. But this is influenced by other factors.

Birds that are heavier, and older, will lay heavier eggs. But birds that are affected by heat, stress and overcrowding will lay smaller eggs. A diet that is low in proteins and fats will also lead to smaller eggs.

CAN BAD EGGS FLOAT?

As we've seen, the eggshell has 7000 to 17 000 tiny pores on its surface. The egg can absorb flavours through these pores. It's an old trick of French chefs to store eggs in a sealed container with some truffles. The cooked egg picks up the flavour of the truffle — and you don't have to use up this expensive fungal delicacy.

It's these pores that make a bad egg float. Bacteria can sometimes enter through these tiny pores and multiply. They will, as a by-product of normal metabolism, generate some gas. There is sulphur in the yolk, so the gas often has some hydrogen sulphide — the classic "rotten egg gas" smell. The gas increases the pressure inside the egg. In fact, one Triple J listener emailed me and said that he had thrown out some bad eggs in the compost. He came back to them a week later and touched them gently — and they exploded because of the excess pressure inside. Not only does this extra pressure inside push out the gas, but it will also push out some liquid.

This is how the egg gets to be lighter — by losing gas and liquids through the pores. After all, both the gas and the liquid have mass. So if you have two eggs of the same volume, and one is lighter, then that's the one that will float. And that's the one that is more likely to be bad.

In some cases, a bad egg might lose only a small amount of gas. In that case, it would still be heavy and would sink. So if you are suspicious about an egg, give it a big sniff for the characteristic rotten egg gas smell.

CAN OLD EGGS FLOAT?

Yup, they can — just like bad eggs. Sometimes life is confusing, and there are no easy answers.

A very fresh egg has a very small air pocket, and will sink and rest in a horizontal position.

As it ages it loses water, and the air pocket gets bigger. It will sink, but with the air pocket at the blunt end uppermost.

By the time it is quite old, it has lost so much water that the air pocket is quite large — and it now floats.

EGG YOLK AND COOKING

Egg yolks are about 50% solid. More than 60% of that solid portion consists of minute globules of fat surrounded by proteins. Yolks are used in baking for their texture, flavour and colour. Egg yolk is also used to thicken soups and sauces.

Egg yolks are the basis of one of the two main principles in making sauces — the Emulsion Principle. Here the egg yolks absorb and suspend particles. The three sauces based on the Emulsion Principle are mayonnaise, hollandaise and vinaigrette (also called French dressing). The other principle is the Roux Principle (which starts with blending a mixture of flour and butter).

BOILED EGGS CAN KILL

And now, an important health message — Boiled Eggs Can Kill!

An egg is a life capsule for tomorrow's chook, and so it has various defence systems.

One defence system includes the hard shell and the soft membranes underneath it. Each of these physical layers can stop bacteria from getting in.

The egg's enzymes make up another defence system. You've probably heard advertisements telling you that certain washing powders contain enzymes, which can eat the stains out of the clothing you wear. The shell membrane and the egg white also contain enzymes, which can destroy invading bacteria.

When you boil an egg, you weaken its defence systems. Bacteria now find it easier to sneak in through the 7000 to 17 000 pores. The heat of the cooking process weakens the enzymes, so they're no real defence against the bacteria.

Luckily, there are usually no bacteria inside the egg, and you usually eat the egg immediately, so there's no problem.

But suppose you cool the boiled egg in water — and then store it for a few days. All water carries bacteria, unless it's distilled water. The water cools the egg down so quickly that a vacuum is formed between the egg white and the shell membrane. This vacuum can "suck" water, and the bacteria in the water, into the egg. The insides of the egg make a superb food supply for the bacteria to grow and multiply.

So this is the take-home message: if you're cooking eggs to store and eat later, cool them down in air, not in water. Then store them in the refrigerator.

Eggs ain't eggs, if they're water-cooled ...

BOILED EGG GETS GREEN RING?

Why, when you boil an egg for a really long time at a high heat, do you sometimes get a green ring around the yolk? The short answer is iron (ferrous) sulphide.

First, if you want to be really accurate, you don't boil the egg. You boil the water, which heats the egg.

Second, the yolk contains iron. It's attached to a coiled-up protein called "phosvitin". The longer that the egg is heated, the more that this protein uncoils — and as it uncoils, it releases the iron.

Third, the egg white contains hydrogen and sulphur in its proteins. Under prolonged high heat the proteins unfold, releasing hydrogen and sulphur. They combine to make hydrogen sulphide gas.

As the egg cools, the solid yolk separates a little from the solid egg white. It's at this boundary that the hydrogen sulphide reacts with the liberated iron to make iron sulphide, in a green layer around the yolk.

There is another effect as well. Sometimes, the green layer is caused by high levels of iron in the water.

And there is yet another effect. Older eggs have yolks that are slightly more alkaline than younger eggs. This alkalinity releases more of the iron from the yolk.

Regardless of how it happens, the iron sulphide is harmless and doesn't affect the taste. But it does look unappealing.

It's easy to prevent. First, don't overheat eggs when you cook them. Second, if you do come across an overheated egg, plunge it straight into cold water and eat it promptly. The hydrogen sulphide will tend to gravitate to the colder shell, away from the yolk.

One last thing. If you sniff deeply on a hard-boiled egg, you can just pick up the very faint odour of hydrogen sulphide. It comes from the normal heating of the proteins in the egg white.

EGG CANDLING

Candling an egg means looking at it in a darkened room while passing it over a strong light. It's an ancient craft.

In the old days, an expert candler could check how the embryo was growing in a fertilised egg. But today, candling is more concerned with quality control. A candler can see abnormalities such as cracks and blood spots.

HEN CHANGES SEX . . .

Sometimes changes in the gonads of hens can turn them into cocks.

In hens, the right ovary has already degenerated within a few days of hatching. But there is still some tissue left in that right ovary. During the hen's active reproductive life, her female sex hormones from her left ovary stop that tissue from functioning.

But as the hen gets older, her production of female sex hormones may slow to virtually zero. When that happens, the remnants of the right ovary can occasionally turn into a testicle. Sometimes, the hen will develop the male plumage, and may start behaving like a rooster.

And sometimes, the hen-turned-to-cock may be able to fertilise other females.

BROODING EGGS

Birds usually keep their eggs at the temperature needed to develop the embryo by sitting on them. As a result, they have evolved a special area of the lower abdomen called the "brood patch". The skin in the brood patch has lost its feathers, and is also very thick and full of blood vessels. These adaptations make a brood patch superb for transferring heat from the adult to the egg. It's usually the female that has the brood patch.

The phrase "brooding hen" refers to a bird that is incubating the eggs and later keeping the baby chickens warm.

References

Peter Aldhous, "Shells give chicks an early break", *New Scientist*, No. 2008, 16 December 1995, p 18.

R. W. Burley and D. V. Vadehra, *The Avian Egg: Chemistry and Biology*, Wiley, New York, 1989.

Anne Gardiner, Sue Wilson and The Exploratorium, *The Inquisitive Cook*, Henry Holt & Company, New York, pp 34–53.

Fred Kern Jr, "Normal plasma cholesterol in an 88-year-old man who eats 25 eggs a day", *New England Journal of Medicine*, Vol. 324, 28 March 1991, pp 896–899.

McGraw-Hill Multimedia Encyclopedia of Science & Technology (CD-ROM), McGraw-Hill, 2000.

Richard Nickel et al, *Anatomy of the Domestic Birds*, translated by W.G. Siller and P.A.L. Wight, Verlag Paul Parey, Berlin, 1977, pp 70–84.

Chilli — BURN BABY BURN

What's a chilli got that makes it so gosh darn hot?

The Cora Indians of western Mexico have a legend about Narama, the First Man. They say that at a feast, he leapt onto a dinner table laden with food. To his surprise, his testicles turned into chilli pods. Narama played it cool, and simply shook his chilli pods all over the food — and that's how humans discovered the hot effects of chilli.

Back in the 17th century, it's mentioned in the tongue twister "*Peter Piper picked a peck of pickled peppers*". The Royal Academy of Dramatic Art used it to improve the articulation and diction of students. And in an episode of "The Simpsons", Homer Simpson had some industrial-grade hallucinations after gorging on a very hot chilli.

So remember: no matter how tough (or stupid) you are, one day you will meet the chilli that you cannot eat.

We still don't understand all the science behind chillies. But we are learning how to re-use them in medicine.

HISTORY OF CHILLI

The history of chillies began about 9000 years ago. That's when they were first cultivated, and used in cooking, in Mexico and Central America.

In 1492, Christopher Columbus left Spain and sailed west in search of India, and the gold and spice islands of the East. He wanted to break the monopoly on pepper and other spices. Instead, he "discovered" Cuba and Haiti, and came across a whole bunch of unfamiliar foods. They included pineapple, maize, sweet potato and our special friend, the chilli. He brought them back to Spain in 1493. They spread rapidly across Europe. In 1525, the Portuguese took chilli peppers to India. In 1894, William Gebhardt from New Braunfels in Texas sold the first commercial chilli powders. He dried capsicum pieces, and then ground them three times in a domestic meat grinder.

BOTANY OF CHILLI

Chillies come from the Solanaceae (nightshade) family. This family includes many important foods, such as the potato, tomato, eggplant and sweet potato. It also includes drugs such as belladonna, and tobacco, which is the world's most economically important drug plant. The Solanaceae family contains the genus *Capsicum*, which includes *Capsicum annuum* and *Capsicum frutescens*.

Chillies belong to the *Capsicum* genus. Capsicums are fleshy-fruited peppers, usually grown as herbaceous annuals. The fruit is a berry, with many internal seeds. The word "capsicum" comes either from the Latin ("capsa" meaning "box", from its pod-like fruit) or from the Greek ("kaptein" meaning "to bite" or "to gulp down"). *Capsicum frutescens* includes the bush red pepper, which gives us the hot Tabasco sauce.

Capsicum annuum has two main sub-families, *Capsicum annuum grossum* and *Capsicum annuum longum*. In the *grossum* sub-family, the capsicums tend to be larger — such as the globular, mild-flavoured sweet peppers. The *longum* sub-family, which includes hot chillies and cayenne peppers, has smaller fruit.

FOR THE BIRDS

Plants and animals gradually evolve survival mechanisms to fight off attackers, get them more babies or help them live better. So why did chillies develop a burning chemical that would deter mammals from eating them and spreading their seeds around, but would still attract birds?

The answer is that there is a big difference between the gut of a mammal and the gut of a bird. There is something in the animal gut that totally stops the chilli seeds from germinating. But when chilli seeds pass through a bird's gut, they germinate. In fact, they germinate better than chilli seeds planted by hand.

There's another advantage. Birds tend to excrete chilli seeds while they are feeding on another bush. So the chilli seeds get planted in the shade — which is exactly the environment they prefer.

CHEMISTRY OF CHILLI

There are several active ingredients that make chillies "hot". They're called "capsaicinoids". J.C. Thresh isolated the first one, capsaicin, back in 1876. Capsaicinoids are found on the seeds, and in fragile glands on the internal white partitions, or ribs, of the fruit or berry. The capsaicinoids tend to be very soluble in fats and alcohol, but very insoluble in water.

> **SIZE OF NERVES**
>
> Nerve cells come in many different sizes. As they get bigger they carry information faster.
>
> The pain nerves involved in capsaicin are called C-nerve fibres. Nobody really knows why other nerves aren't involved. They're skinny and carry non-urgent information slowly to the brain. A-nerve fibres are larger, and they can carry information more quickly.

WHERE THE HOTNESS HAPPENS

In the body, chemicals often work by landing on a cell receptor that is specially tuned to accept that chemical. In 1997, scientists finally discovered where the capsaicinoids do their hot thing. Michael J. Caterina and his colleagues from the Department of Cellular and Molecular Pharmacology, Anesthesia, and Medicine at the University of California in San Francisco were the first to isolate the capsaicin receptor. They called it VR1, for Vanilloid Receptor Subtype 1.

Why do chillies burn?

The chilli had its "cool" beginnings around 9000 years ago in Mexico and Central America.

Chemicals often work by landing on a cell receptor that is specifically tuned to respond to that chemical. For hot stuff, this is the capsaicin receptor, found mostly in the mouth.

MORE CHEMISTRY OF CHILLIES

There are many active ingredients in chillies. Thresh isolated the first one, capsaicin, in 1876. It was thought to be the only active ingredient until 1960. Since then, we've discovered several other active ingredients. One chemical called di-hydrocapsaicin is as active, or as "hot", as capsaicin. All the other capsaicinoids discovered so far are less fiery.

Capsaicin is a crystalline solid with the formula $C_{18}H_{27}NO_3$. It melts at 65°C. The chemical name for capsaicin is 8-methyl-n-vanillyl-6-nonenamide. Its structure was finally identified in 1984.

In the chillies of the *Capsicum annuum* species, there are roughly equal amounts of capsaicin and di-hydrocapsaicin, making up between 0.1% and 1% of the total weight. In *Capsicum frutescens*, the active ingredients are present in concentrations of 0.4–1%, but with twice as much capsaicin as di-hydrocapsaicin.

This capsaicin receptor sits on the cell membrane of pain nerves (called "nociceptors") on the inside of the mouth. When capsaicin lands on the capsaicin receptor, it forces it to open up, which lets calcium and sodium ions (in a 10:1 ratio) rush into the cell. This massive influx stimulates the pain nerve into firing, which sends a signal to the dorsal root ganglion in the spine, and finally to the brain. Your brain then turns this signal into sensations of pain and heat.

The capsaicin receptor can also be fired up by other stimuli such as other chemicals, or mechanical or thermal stress. For example, inflicting a sudden temperature change from 22°C to 48°C makes the receptor fire.

Another chemical, called capsazepine, blocks the action of capsaicin. Perhaps restaurants that specialise in very hot chillies might consider having capsazepine handy for their more foolhardy customers.

My 11-year-old son and I once tried to make, and eat, successively hotter curries. We gave up when we got to a nuclear-grade curry that was so hot that we both temporarily lost all sensation on small symmetrical areas of our faces near the chin ...

The mouth has a very dense network of nerve endings, so capsaicin can cause thermonuclear-grade heat there. But there are fewer nerve endings in the oesophagus. This might explain why the nuclear fire cools down to mere simmering coals as your favourite chilli slides down towards your stomach.

HOW TO COOL THE BURN

Back in 1990, Christina Wu Nasrawi and Rose Marie Pangborn (from the Department of Food Science and Technology at the University of California at Davis) wrote an interesting paper called "Temporal effectiveness of mouth-rinsing on capsaicin mouth-burn". They tested different "cooling" liquids, at different temperatures. The liquids were plain water, milk, sugar-and-water and alcohol-and-water. Their 22 brave unpaid volunteers were aged between 22 and 45.

Cool the Burn — 1

One result stood out. Cold liquids at 5ºC were more effective in neutralising the chilli mouth-burn than warmer liquids at 20ºC.

We're not sure what is going on. Maybe the "cold-sensitive" nerves inhibit the "heat-sensitive" nerves from sending their signals. Maybe the coldness slows the "heat-sensitive" nerves so much that they don't carry signals any more. We really don't understand.

Cool the Burn — 2

Another result was that a 10%-sugar solution of water at 20ºC is just as effective as cold milk at 5ºC. That's also a little confusing.

We know that, in the lab, the capsaicinoids dissolve easily in fat. In your mouth, the fat in milk supposedly dissolves the capsaicin, and removes it from the VR1 receptor. That's why milk relieves the burning sensation of chilli — plus it's cold. That makes perfect sense.

But why should a sugar-and-water solution that is 15ºC warmer be as soothing as milk? After all, capsaicin doesn't dissolve in water. Again, we don't know (see box, *Sugar and Mouth-burn*).

Cool the Burn — 3

Another strange result was that a 5%-alcohol solution is no better than plain water in relieving the hot sensation of chilli. We know that capsaicin is (in the lab) very soluble in alcohol. So alcohol should be more soothing than water. Perhaps alcohol's "irritant" effect in the mouth is greater than its "soothing" effect of dissolving and removing the capsaicin.

These experiments tell us that we don't fully understand how chilli causes mouth-burn. There's much to discover.

Cool the Burn — 4

Another sure-fire way to reduce the burn of a chilli is to vigorously move your mouth and tongue.

This movement stimulates the mechanoreceptors on nerves in your mouth. These activated mechanoreceptors seem to inhibit your pain nerves.

Cool the Burn — Summary

So if you accidentally eat a very hot chilli, quickly drink something cold (either milky or sugary). Once you've swallowed it, move your tongue and mouth. And avoid alcohol. Ice cream seems to be an ideal cure . . .

SUGAR AND MOUTH-BURN

Sugar can kill pain. Studies with babies show that a sugar-and-water solution can reduce the pain of needle vaccinations. If you give them sugar before a vaccination, they cry for a shorter time than babies who missed out on the sugar. Sugar seems to activate the pain-killing endorphin chemicals in babies.

Perhaps endorphins also get released when adults drink something sugary. Maybe that's why sweet drinks can relieve the pain of a particularly fiery chilli.

SCOVILLE UNITS = TASTE RICHTER SCALE

Back in 1912, Wilbur L. Scoville, a pharmacologist working with Parke-Davis, came up with a test to measure the "heat" of a chilli.

There is an excellent test for measuring very small quantities of chemicals in liquids: High Pressure Liquid Chromatography. Unfortunately, this test does not measure what humans feel. But the Scoville Organoleptic Test does.

Dr Scoville dissolved one grain (0.0648 grams) of raw hot pepper in 100 cc of alcohol and let it sit overnight. He then put a small drop of this liquid on a volunteer's tongue. If the volunteer could taste it, Scoville would dilute the alcohol with water and test it on the tongue again. He would continue diluting and testing until there was no sensation of heat. The chemists call this an Extinction Test, or a Dilution Taste Procedure. If a single drop of this alcohol-and-pepper solution had to be diluted with 30 drops of water, then that pepper would rate at 30 Scoville Units.

The Scoville Test is only approximate, because it doesn't take account of natural variation in plants from season to season. Also, it's a subjective psychophysical scale — ie, it depends on a person's opinion. Various groups (such as the American Spice Trade Association and the International Organisation for Standardisation) have suggested different testing methods — but Scoville Units are still very popular.

The Scoville Scale runs from 1 up to about 16 million.

Chilli Fruit	**Scoville Units**
Commercial taco sauces	300
Poblano	1000–1500
Jalapeno, chipotle, guajillo	2000–5000
Serrano	1000–23 000
Cayenne pepper and tabasco	30 000–50 000
Thai "mouse dropping" peppers	50 000–100 000
Charleston Hot	100 000
Scotch Bonnet	100 000–300 000
Habañero	200 000–300 000
Red Savina	325 000–577 000
Naga Jolokia	855 000

Capsaicinoid Chemicals*	**Scoville Units**
Homocapsaicin	8.6 million
Homodihydrocapsaicin	8.6 million
Nordihydrocapsaicin	9.1 million
Capsaicin/Di-hydrocapsaicin	16 million

*See box, *More Chemistry of Chillies*.

CHILLI — THE WEAPON

Chillies don't bother birds, because they don't have the VR1 capsaicin receptor on their cell membranes. Birds happily eat chillies and spread the seeds around the countryside. But capsaicin really upsets mammals, which includes us humans.

Indian women have long used powdered chilli as a weapon to throw in the eyes of attackers. In southern Andhra Pradesh, female temperance squads who found men drinking alcohol would hit them with brooms, hold them down and force-feed them chilli powder.

The modern version of this is the capsicum spray used by police officers. It carries up to 5% red pepper dissolved in natural oils. The spray is most effective at 2 metres, but it has a range of up to 4 metres. One effect is to immediately irritate and engorge the tiny blood vessels in the eye. This causes a temporary blindness lasting five minutes. The other thing that the capsicum spray does is inflame the membranes in the nose and throat, which makes it very difficult to breathe for about two minutes.

The capsicum spray is supposedly a non-lethal spray to be used only for self-defence. But if you take in enough, chillies can be lethal. If you weigh less than 63.5 kg and eat 2.84 litres of a Louisiana-style hot sauce, you have a very high probability of dying from respiratory failure. However, most people would find this almost impossible to do.

CHILLIES MAKE YOU FAT

Chillies can make you eat more. Your stomach has stretch receptors, which monitor how full it is. These stretch receptors are connected to nerves, which tell your brain when your stomach is full.

Chilli in your stomach can inhibit these nerves. Eat chilli, and you might eat more.

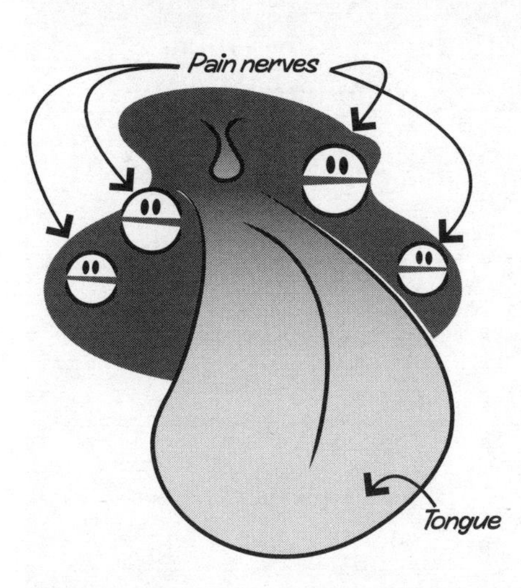

Capsaicin receptors sit in the cell membranes of pain nerves ("nociceptors") on the inside of the mouth.

Capsaicin — the hot stuff in chilli — forces the receptor to open up, which lets calcium and sodium ions rush into the cell and stimulate the pain nerve into firing HOT signals.

CHILLI PREVENTS STOMACH ULCERS?

Dr Khay Guan Yeoh, from the National University Hospital in Singapore, has found that chilli dishes could give protection against stomach ulcers.

He and his team worked with 18 healthy volunteers. They gave them 600 mg of aspirin, which can produce temporary ulcers in the stomach. They used a gastroscope to look at the volunteers' stomachs both before and after the aspirin.

Without the chilli, 600 mg of aspirin gave the volunteers four little ulcers on average. But if the volunteers had eaten 20 grams of chilli powder 30 minutes before they had the aspirin, on average they had only 1.5 ulcers. The chilli powder protected them against stomach ulcers.

The scientists aren't sure how the chilli powder works. One theory claims the chilli might work by increasing blood flow to the mucosal lining of the stomach. In turn, this could force the mucosal lining to make protective chemicals called prostaglandins.

CHILLI AND CANCER PAIN

Another study has shown that chilli can reduce pain in cancer patients.

One common approach to treating cancer is "slash and burn". The surgeon "slashes" the cancer with a scalpel to remove as much of it as possible. Then the oncologists "burn" what is left with drugs, or radiation, or both.

The drugs (chemotherapy) kill the fast-growing cells in the cancer — and everywhere else in the body, including in the mouth. In fact, 40–70% of cancer patients receiving chemotherapy complain of a sore mouth. The pain can be bad enough to interfere with speaking and eating.

CHILLI AND BIG ANIMALS

Jack Birochak from Valley Forge in Pennsylvania has designed and built pepper sprays powerful enough to repel grizzly bears. A typical spray-can holds about 1 kg of chilli pepper and oil. Now he's been asked to develop an industrial-grade version to repel elephants.

In Asia and Africa, elephants destroy lots of crops. In Zimbabwe, about 100 elephants are killed each year in the process of stopping them from attacking crops. The elephant has one of the most sensitive senses of smell in the animal kingdom, thanks to its very long trunk lined with mucous membranes. So chillies could work.

But it's risky to get close to a big elephant and threaten it. So Birochak is working on a compressed air launcher to toss the can of pepper spray some 200 metres through the air. It can be set to start spraying either on impact, or in mid-air.

Dr Ann Berger of the Yale Cancer Centre, and Professor Linda Bartoshuk from the Ear, Nose and Throat section of the Department of Surgery, also at Yale, have done research into pain relief. They added cayenne pepper to a soft chewy toffee, giving it a 100 Scoville Unit rating.

The capsaicin overstimulated branches of the trigeminal nerve in the mouth, leaving it temporarily unable to respond to other pain stimuli. This treatment was superior to other pain-relieving treatments such as anaesthetics or narcotics, which made it difficult for the cancer patients to chew. The patients were asked to rate their mouth pain from one to ten (least to most). On average, the capsaicin candy reduced the intensity of the mouth pain from a rating of 6 to only 1.5 — and this improvement lasted for several hours.

At the time, most of the local male doctors dismissed Doctors Berger and Bartoshuk as "*a couple of broads making candy*". But other scientists are now looking at using chilli derivatives to treat chronic pain caused by such conditions as arthritis, shingles, diabetic neuropathy and spinal-cord injuries.

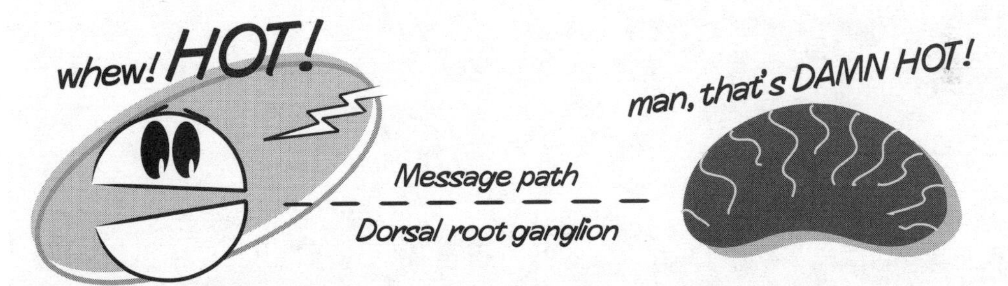

After stimulation, the pain nerve sends a signal via the dorsal root ganglion (located in the spine) to the brain, where it turns into sensations of pain and/or heat!

ODD USES FOR CHILLI

The Cuna Indians of Panama would trail a string of chillies behind their canoes when they went to sea. This was supposed to discourage sharks from attacking.

Another novel use for chilli is as an anti-fouling agent on boat hulls and other underwater structures. Barnacles increase the drag on a moving boat, and can increase the fuel bill by 30%. The hot chilli apparently repels any barnacles, mussels and tubeworms that want to take up residence.

Chilli has also been used on underground cables to stop rodents from chewing on them, to repel cats and dogs, and to encourage people not to bite their fingernails. Capsaicin derivatives are also used in hair and dandruff preparations, and in toothpastes to *"keep the mouth refreshed and free from bad breath and fur"*.

CHILLI AND OTHER PAIN

Capsaicin is also being used to treat patients with hypersensitive and hyper-reflexic bladders. Patients with this condition complain of pain in the bladder, the sudden onset of the need to urinate and a very high frequency of such a need.

The study involved squirting very small quantities of capsaicin directly into the bladder. This treatment reduced the pain, frequency and urgency for up to 16 days — and then the symptoms gradually returned.

The capsaicin probably reduced the sensitivity of the nerves in the bladder wall.

And finally, an ancient Coptic remedy reads: "*Someone with an itching anus: Knead the honey with burnt wolf's dung ground with white pepper. Let the patient drink it. But first claim your fee.*"

CHILLI FOR EXERCISE

One small study found that athletes who ate chillies before exercise burned carbohydrates more efficiently. Furthermore, they performed better than the athletes who did not eat chillies before their exercise.

Unfortunately, the numbers were not really big enough to give solid statistics. This study was carried out on only eight middle-distance and long-distance male runners. It would be interesting to see what a study with bigger numbers would show.

Chillies have an overall thermogenic effect. This means that eating chillies makes you burn up extra carbohydrates, even if you don't exercise.

The trick is not to over-eat (see box, *Chillies Make You Fat*).

References

Michael Caterina et al, "The capsaicin receptor: A heat-activated ion channel in the pain pathway", *Nature*, Vol. 389, 23 October 1997, pp ix, 816–824.

Henry Fountain, "Survival of the hottest", *The New York Times*, 31 July 2001.

Thomas Levenson, "Studies from life", *The Sciences*, January–February 1995, pp 13–15.

C.W. Nasrawi and R.M. Pangborn, "Temporal effectiveness of mouth-rinsing on capsaicin mouth-burn", *Physiology & Behavior*, Vol. 47, No. 4, April 1990, pp 617–623.

Joshua J. Tewksbury and Gary P. Nabhan, "Directed deterrence by capsaicin in chillies", *Nature*, Vol. 412, 26 July 2001, pp 403–404.

John Travis, "Hot stuff: A receptor for spicy foods", *Science News*, Vol. 152, 8 November 1997, p 297.

Fraudulent FIREWALKING

Is it really possible to walk on hot coals without burning your feet — or your wallet?

On a really hot day when the air temperature rockets to 40°C, the Sun-soaked asphalt of your average two-lane black-top road can reach 68°C. But temperatures much less than 68°C can easily put you in hospital. A temperature of "only" 44°C for as little as 35 seconds can give your bare feet second-degree burns, according to doctors at the Maricopa Medical Center in Phoenix, Arizona. (Each summer, when the hot Arizona Sun forges the Streets of Fire, the doctors at the Center see about three or four people with "pavement burns".)

So how can firewalkers possibly walk on hot coals at over 600°C with no ill effects?

A SHORT HISTORY OF FIREWALKING

The history of firewalking goes back to 1220 BC. At various times in history firewalkers have strolled over hot coals in Greece, Spain, India, Bulgaria, Fiji and Sri Lanka.

In the old days, firewalking had a strong religious or spiritual component. It was often part of a purification and healing ritual. If you succeeded in crossing the hot coals without getting burnt, that was because mysterious spiritual and/or mystical powers had protected you. If you failed and got burnt, you obviously hadn't reached a high enough plane of consciousness.

Today, there are many organisations that are happy to exploit your personal insecurities and separate you from your money. But if you want to firewalk without burns, you don't have to pay large sums of money to motivational speakers who supposedly empower you to get in touch with your inner self.

The Modern Age of Firewalking-for-Money began in 1977, when Tolly Burkan in California had his first

"FIREWALKER"

Actually, the name "firewalker" is not strictly accurate. They don't walk on fire. In fact, there's no flame at all involved. They don't even walk on naked hot coals.

They walk on hot coals which are covered by ash — but "ashwalker" doesn't sound as impressive as "firewalker".

firewalk. By the early 1980s, he had set-up his Firewalking Institute of Research and Education — FIRE. Very rapidly, interest in firewalking snowballed. Tolly Burkan claims that millions of people have now experienced firewalking. Indeed, one of his FIRE-certified instructors, Bill Bastian, did a firewalk on an episode of *Survivor*.

By 2000, Tolly Burkan had moved away from providing commercial firewalking to the general public. But he still offers to the very lucrative corporate market a product he calls High Tech Fire — a "*high-powered streamlined seminar that delivers explosive transformation in two hours*". He says that High Tech Fire "*is changing the consciousness of corporate America ... stimulates greater resourcefulness and creativity in management*". By way of explanation, he draws on neuroendocrinology (nerves and hormones) and claims that "*thoughts change brain chemistry ... resulting in alteration of body chemistry as well*".

FIREWALKING 101

One undeniable advantage of firewalking is that it improves your self-confidence — but then, so does success in any physical feat.

Firewalking has nothing to do with mind balance, ESP, faith, moral certitude, a mystical appeal to higher entities, altered states of consciousness, psychic energies, psychological preconditioning, a mysterious and undefinable spiritual force of bioenergy, regenerative potential, realignment of your chakras or kundalinis, neurolinguistic programming, fasting, celibacy, devotional chants, mental syntax, frequent baths, no baths, quantum mechanics, mind-over-matter, or even paying money for the privilege.

Firewalking can be explained, totally and simply, by the laws of physics — with a little physiology thrown in. The simple truth is that practically anybody can firewalk safely — all you need is agile feet.

HOW IT WORKS — EXAMPLE

Imagine that you have a cake cooking in a cake tin inside an oven at 180ºC. It has been baking for an hour, so everything in the oven is at 180ºC.

You open the oven and plunge your hand into the hot air at 180ºC. Your hand doesn't get burnt by the hot air. You gently, with your naked finger, touch the top of the cake, which is also at a temperature of 180ºC. Once again your finger doesn't get burnt. But if you touch the hot cake tin, you'll get large blisters on your naked fingers. So you have to grab an oven mitt before you can remove the hot tin and cake.

Why do the air, cake and metal cake tin all have different abilities to burn you?

HOW IT WORKS — PHYSICS

It's all to do with two separate "quantities" — heat content, and heat conductivity.

Heat content measures how much heat energy an object can store. Air is very thin, and doesn't have a lot of mass, so it can't store much heat energy. But both the cake and the cake tin are more dense, so they can store lots of heat energy.

However, heat conductivity means that only the metal cake tin burns your fingers. Heat conductivity measures how rapidly heat energy can flow out of an object. "Conductivity" and "insulation" are opposites. If an object is a good conductor, it's a bad insulator — and vice versa.

Air is a bad conductor and a good insulator, so you can "touch" the hot air in the hot oven and not get burnt. The same goes for the cake. Even though the cake has a lot of heat energy stored in it, the poor conductivity stops the heat from getting into your hands.

But the cake tin has both a high heat content and a high conductivity. Touch the tin and you'll get burnt.

FIREWALKING RECORDS

David Willey teaches Physics at the University of Pittsburgh in Johnstown. He has demonstrated various principles of physics by striding across broken glass, having concrete blocks smashed on his chest while lying on a bed of nails, plunging his naked hand into a tub of molten lead, picking up glowing amber Space Shuttle tiles with his naked fingers, and yes, walking across hot coals with his bare feet.

He is totally sceptical of any need for spiritual preparation before any of these experiments.

He collected the data for the "hottest fire intentionally walked on by a human being". On 18 October 1997, near Redmond in Washington State in the USA, Michael McDermott walked across a fire bed some 3.5 metres long. The hottest part of this bed, a patch about 50 cm long, ranged in temperature from 872°C to 989°C.

He also holds the record for the "longest firewalk". He did this on 2 July 1998, on his home campus. He and 14 others successfully walked along a firebed 50.3 metres long.

The worst injury suffered during any of his walks has been a blister smaller than a fingernail.

Let the fire force be with you

Step up and **PAY ME** for the spooky secrets.

OR

Physics 101

It's all to do with **TWO** separate "quantities"...

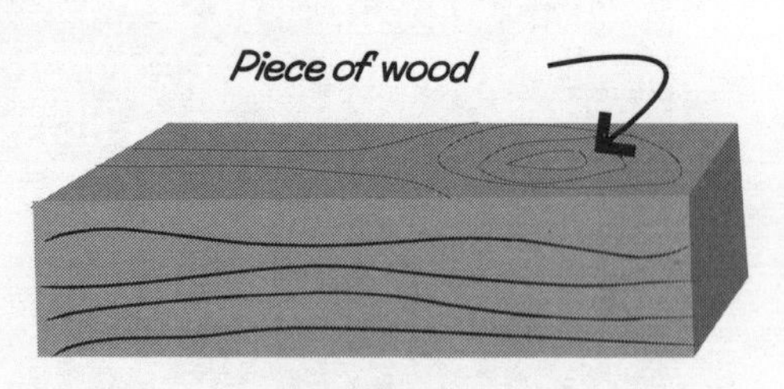

... Wood is a poor conductor of heat ...

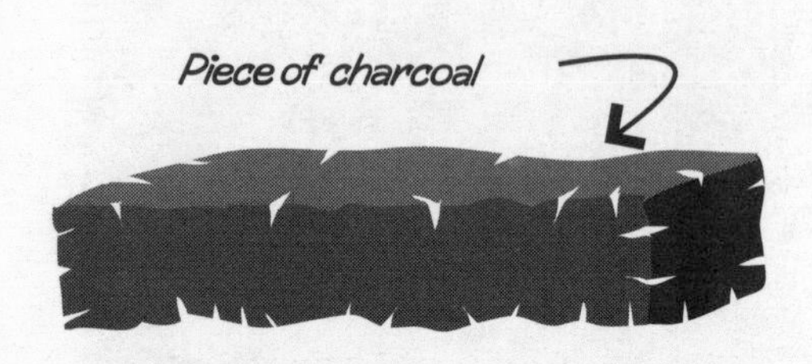

... Charcoal is an even worse conductor, which makes it an excellent insulator!

... When aiming to impress good folk by walking on something hot, charcoal is the obvious answer!

HOW IT WORKS — PHYSICS 2

There's another factor helping you walk across hot coals — the surface of the coals is a lot cooler than the inside.

Heat is transmitted by three main methods: conductivity, convection and radiation.

A good example of conductivity is a long bar of metal in a hot fire — with one end in the hot coals, and the other end a metre away, out in the cold. The heat energy will "conduct" up along the metal bar, bouncing from atom to atom. After a while, the end that was cold will start to warm up.

You can see convection every time you boil water in a saucepan on your stove. The water nearest the flame (at the bottom of the saucepan) will get hot, become less dense, and then rise to the top.

This is convection — where the molecules that are hot actually travel from one place to another.

Radiation is what you feel when the Sun shines on you through a glass window in winter. Radiation even works through a vacuum — which is how the Sun heats up the Earth.

The hot coals at a firewalk have a large surface area. They immediately begin to lose their heat through radiation. The surface rapidly begins to cool. But the coals are magnificent insulators — which means poor conductors. They can't quickly transfer the heat energy to the surface. So the surface stays cool.

WOOD IS GOOD . . .

Wood is a very good insulator against heat — and a very poor conductor of heat. That's why wood has been used for handles on frying pans for centuries. You can also burn wood to get charcoal.

CHARCOAL IS BETTER . . .

Charcoal is four times better as an insulator than wood. It's several thousand times better as an insulator than metal. And charcoal has a very low heat content. If you want to impress people by walking on something hot, charcoal is the obvious answer.

It's not a higher state of spiritual awareness that protects you from blisters — it's basic high school thermo-dynamics. If somebody tries to take your money for firewalking, ask them to prove that they have something really special to offer. Ask them to walk on a hot steel plate!

1930S INVESTIGATION

Back in the 1930s, the Council for Psychical Research at the University of London organised two separate firewalks.

The first one, in 1935, had two British scientists and an Indian, Kuda Bux, walking across a fire pit 3.6 metres long. The embers came from oak, and were at a temperature of around 430ºC. Chas R. Darling reported on this firewalk in *Nature*. He wrote: "*the heat was merely another form of the fireside experiment of picking up a hot cinder and returning it to the fire, when the fingers are not burnt if the action is performed quickly*".

Their second firewalk was in April 1937, again by several people including an Englishman, Reginald Adcock, a Muslim, Ahmed Hussain, and several others.

In neither firewalk did the participants engage in any form of spiritual purification, religious enlightenment, or any other ritual preparation. Even so, there were no significant injuries.

So the Council for Psychical Research announced that the secret of the firewalk was quick walking and plain physics — a combination of the incredibly low thermal conductivity and low heat content of the coals, and the short time of contact between the coals and the feet.

MODERN INVESTIGATION

Back in 1997, Kjetil Kjernsmo, a Norwegian scientist from the University of Oslo, investigated the phenomenon of firewalking with heat-sensitive cameras.

The first thing he found was that the temperature of the coal bed was very uneven. His camera saw temperatures across the coal bed varying between 150ºC and 700ºC — 700ºC is hot, but nowhere near the 1200ºC that some firewalkers claim to have measured.

The second thing he found was that very little heat energy left the coals and entered the naked feet during a firewalk. According to his heat cameras, the naked feet didn't cool down the coals much at all — very little heat energy left the coals. And on the receiving end, after a typical firewalk, the bare soles heated up by only 4Cº.

Another factor that makes firewalking just a matter of physics is that the surface is usually very uneven. This means that there's a very small surface area of your foot actually touching the hot coals at any given time.

Finally there is some physiology involved. The outer layer of human skin is dead. Even people who always wear shoes and have soft feet have enough dead skin to provide good insulation. And as a further means of assistance, blood circulating through the feet is an excellent conductor, and helps to take the heat away.

HOW FIREWALKING WORKS

Wood coals are always used, never BBQ coals, which are quite different. And care

FIREWALK WITH A STEAK

Here's a simple experiment to test how much heat energy comes out of a bed of coals by direct contact or touch. The physicist Bernard Leikind did this experiment in 1994.

He turned up at a firewalk with two sirloin steaks and strapped one to the bottom of each foot. He was able to walk across the coals and saw that the steaks were unaffected.

He then placed a metal grill (which conducts heat very well) on the coals. He gave the grill time to heat up by radiation (through the air), so that it was at the same temperature as the coals. When he put the steaks on the metal grill, they instantly began to sear.

is taken to make sure there are no nails or other metal in the wood that is used.

First, the coals are raked as smooth as possible, so that they are covered by dead ashes. The dead ashes are an excellent insulator against heat. Ashes are not essential, but they increase the safety margin. If a gust of wind removes the insulating layer of ashes, the fire is smoothed over again with more ashes. Any large burning coals are removed or covered over. To make the firewalk more dramatic, it is usually performed at night, so that the audience can see the hot coals glowing through the dead ashes.

Once the firewalker enters the coal bed, they keep moving steadily. In a typical 3–4 metre firewalk, the total time of contact with the coals will be less than one second for each foot. A firewalker will never run — the pressure of landing can be enough to make a hot coal stick to the skin. The toes are kept curled upwards, so that hot cinders are less likely to get in between them. It also helps if the firewalker's feet have calluses.

Of course firewalking, like any similar physical activity, is never completely safe.

PERSONAL EXPERIENCES

Barry Williams, of the Australian Skeptics, has also walked on hot coals. He definitely had no spiritual or religious preparation. He wrote: "*the surface was quite rough, not unlike a gravel road, and it hurt my feet a little, though not to any great extent. The heat was much less than that I have felt walking barefoot on a bitumen road in mid-summer. In fact, I felt more heat from the radiation from the coals than I did from the direct contact or conduction.*"

Steve Moneghetti, one week after winning the Tokyo Marathon, walked on coals in Ballarat. Besides being a great athlete, Steve is also a teacher of Science and Maths. He described the sensation thus: "*it's not even as hot as walking on hot concrete. It's all to do with heat transference from the charcoal: it takes a while for the heat to be transferred from the coals to your feet. So long as you don't stand still and there is no metal in the pit, you shouldn't get burnt.*"

SHUTTLE TILE WALK?

The tiles from the Space Shuttle are magnificent insulators — among the best ever made by the human race. I have seen footage where Space Shuttle tiles had been put in an oven at over 1000°C. After a few days, they had heated all the way through to over 1000°C.

Then a human hand moved into shot and picked up one of the glowing red-hot tiles with naked fingers. The tile was such a good insulator that hardly any heat energy flowed from the red-hot tile to the fingers.

And of course, thanks to radiation, the surface cooled down very quickly.

THE FULL CIRCLE

In an ironic turn, firewalking has made its way back into the Third World. Denis Dutton, a past president of the New Zealand Skeptics, travelled to New Guinea. While there he trained a local tribe in how to firewalk, so they could better separate tourists from their lovely dollars.

They were a little nervous at first. But eventually the whole tribe, including the little children, performed the Firewalk Ceremony. Of course, they personalised it by adding their own local rituals and magic incantations. Before he left, he asked them how they would explain to visiting tourists their ability to firewalk, when none of the other tribes in the area could.

Their answer was disarmingly simple: "*We'll just say that an alien from the skies came and taught us.*"

WOODEN SPACE SHIPS

Charcoal is such a good insulator that NASA thought about using it as a heat shield in the early days of their Space Program. Their tests showed that while charcoal was one of the best insulators, it was too brittle. So if there was any uneven air flow over the nose of the incoming space ship, the turbulence could rip chunks of charcoal out. The exposed metal that had been under the charcoal would quickly melt. The advantage of the Space Shuttle tiles was that they were less brittle and had greater internal structural integrity, so they could survive a turbulent re-entry.

However, the Chinese have successfully used charcoal for heat insulation on their re-entry vehicles.

References

Phillip Adams, *The Skeptic*, Vol. 5, No. 2, 1985, pp 3–4.

"Burning soles", *New Scientist*, No. 2003, 11 November 1995, p 13.

Chas R. Darling, "Fire-Walking", *Nature*, 28 September 1935, p 521.

William Z. Harrington et al, "Pavement temperature and burns: Streets of fire", *Annals of Emergency Medicine*, Vol. 26, No. 5, November 1995, pp 563–568.

Adam Joseph, "Olympian braves the coals", *The Skeptic*, Vol. 14, No. 1, 1994, pp 6–8.

David Willey, "Firewalking myth vs physics", November 2000 — www.csicop.org/si/9911/willey.html.

The Great Wall of China FROM SPACE

Is it really possible to see the Great Wall of China from space?

There's a myth that you can easily see the Great Wall of China from the Moon. Totally wrong — you can't even see the Great Wall from the Space Shuttle, unless you've had a lot of practice, or use binoculars.

Imagine looking out of a third-storey window and trying to see a very long fishing line on the ground — you can't. It doesn't matter how *long* the fishing line is — it's simply too skinny to be seen. Same with the Great Wall of China.

THE MYTH BEGINS

The American traveller William Edgar Geil wrote a book called *The Great Wall of China* in 1909. In it, he claims that you can see the Great Wall from the Moon.

Then in 1923, Adam Warwick wrote in the *National Geographic*: "*According to astronomers, the only work of man's hands which would be visible to the human eye from the Moon is the Great Wall of China.*"

Joseph Needham went even further — all the way to Mars. He wrote, in 1971 in

Volume 4 of his multi-volume work *Science and Civilisation in China*: "*Stretching from Chinese Turkestan to the Pacific in a line of well over 2,000 miles (nearly a tenth of the Earth's circumference), the Wall has been considered the only work of man which can be picked out by Martian astronomers.*"

In 1981, the book *The Great Wall,* edited by Luo Zewen, had a foreword by Jacques Gernet that read: "*Proclaimed as the only man-made structure that can be seen from the Moon, the Great Wall of China is a subject of astonishment to westerners.*" An author of one of the chapters in this book, Dick Wilson, wrote: "*. . . had not American astronauts flown through space to prove empirically what everyone below had been saying, namely that the Great Wall of China is the only man-made monument visible from the Moon?*"

In 1983, Salen Lindblad Cruising Inc. of New York ran advertisements for their Chinese cruise with the enticement that "*You can see the Great Wall of China from the Moon. Or see the Moon from the Great Wall.*" One such advertisement appeared in the January–February 1983 issue of *Archaeology*.

In 1988, the Royal Viking Line published a similar advertisement in the January 31 issue of the *San Diego Union*. It read: "*. . . it began 400 years before Christ. It is visible from Mars. You can touch it this Spring. Only one man-made landmark can be seen from outer space without a telescope: the Great Wall . . .*"

THE FIRST EMPEROR

In 246 BC the boy king Chao Cheng took control of the Ch'in state at the age of 13. Ch'in was one of many small warring states that made up China between 771 and 221 BC. By 221 BC, Chao Cheng had defeated the other hostile states, and had brought them together under his rule to form a single empire. Chao Cheng called himself "Ch'in Shih huang-ti", which means "First Sovereign Emperor of Ch'in". In fact, the name "China" comes from "Ch'in".

Shih huang-ti was strong, energetic and brutal. He was described as having *"a high-pointed nose, slit eyes, pigeon breast, wolf voice, tiger heart, and stingy, graceless, cringing character"*.

He changed the country.

He brought in standards for the width of highways, writing, and measurements of weight and length. He abolished the feudal system, and ran China with his own officials, whom he sent out across the country to enforce his laws.

He got rid of all books, unless they dealt with law, herbal medicine or horticulture. When Chinese scholars criticised him, he executed them. One legend says that he invited the scholars to speak freely at a huge banquet, listened closely to their criticisms, and then had them killed immediately after dessert. A variation on this legend says that he burnt the classic Chinese books in 213 BC, and when the scholars complained he had them burnt alive. It is said that he had the books burnt so that Chinese history would begin with him.

He did not have a high opinion of scholars. He wrote: *"Men of letters, as a rule, are very ill acquainted with what concerns the government of a country — not that government of pure speculation, which is nothing more than a phantom, vanishing the nearer we approach it, but the practical government, which consists of keeping men within the sphere of their proper duties."*

Shih huang-ti was terrified of death. He tried everything to achieve immortality: he offered sacrifices to the deities, and even sent out explorers to uncover the Elixir of Life. He died in 210 BC, while touring his empire. His tomb is guarded by 6000 life-sized terracotta soldiers. The total funerary compound covers some 50 sq km.

His empire collapsed about four years after his death; after he died his younger brother seized power. The second Emperor did not have his power and determination. He foolishly did not immediately crush a rebellion that broke out. He was killed by a eunuch minister, and the rebels took control in 206 BC.

THE HISTORY OF THE GREAT WALL

The Chinese have been building major walls for at least 2500 years. Chinese chronicles going back to the fifth century BC describe long walls that separated one internal feudal kingdom from another, or protected them against foreign enemies.

But over the last 2500 years, these various walls have been allowed to decay and have later been rebuilt. This cycle has repeated itself many times. Today, some sections are in as-new condition, while others cannot even be recognised as having once been a wall.

EMPEROR SHIH HUANG-TI

In 221 BC, Shih huang-ti came to power after unifying many small states into a single country called China. The area he had to govern was so huge that he built an extensive network of highways to shift his troops around quickly.

To protect his farmers against invading nomads from the Great Steppes to the north, he connected and fortified all the existing walls. His "Long Rampart" stretched from the Po Hai (Gulf of Chihli) in the east, across Inner Mongolia to the edge of Tibet in the west. He strengthened the walls, and added signal towers and garrison stations. The Long Rampart was begun in 219 BC, and finished 15 years later in 204 BC (seven years after his death).

The labour force needed was immense. The workers included some 300 000 soldiers, prisoners-of-war and even criminals from the jails. The distances were so huge that it was extremely difficult to get supplies to the distant reaches of the Rampart. One chronicler wrote that "*of 182 loads of grain dispatched, only one would reach its destination, the rest being eaten or sold along the road*".

One legend explains why the Wall has so many branches. It claims that the Emperor had a magic white horse, which he allowed to wander freely. The builders had to follow it, putting up fortifications wherever it went. According to the legend, "*the workmen … called a halt to drink their tea … After tea they continued in the same line for 10 miles.* [*They eventually*] *… spied the animal far away … heading in quite a different direction. So the workmen abandoned the last stretch, returned to their camp and built a new wall …*"

GREAT WALL?

Western countries call it "The Great Wall of China". But the Koreans call it "The 1000-Mile Castle". The Japanese have a different name. They call it Banrinochoujou, "The 1000-League-Long Castle".

The Chinese themselves call it "The 10 000-Li Castle" or "The 10 000-Li-Long Wall". A "li" is about half a kilometre.

When is a wall not a wall?

It is a myth that you can easily see the Great Wall of China from the Moon ... and don't let anyone tell you different.

In fact, it would be almost impossible to see it from the Space Shuttle, unless you've had a lot of practice (or are using binoculars).

LATER WALL WORK

Some of the Wall was built of stone, but much of it was built of mud, and so it crumbled with time. By the sixth century AD, so much had eroded away that the Tungustic Wei and Tsi Dynasties (386–577) wrote of building, not rebuilding, the Great Wall. They also added a new loop running between Beijing and Kalgan.

But their work also decayed with time.

So the Great Wall was rebuilt yet again during the Ming Dynasty (1368–1644), which had 16 emperors. During their 276-year reign they often had to repel attacks by the barbarians

THE RESOLUTION OF THE HUMAN EYE

How finely do you see? A circle has 360 degrees. An arcminute is 1/60 of a degree. An arcsecond is smaller again. It's 1/60 of an arcminute, or one-3600th of a degree. The maximum resolution of the human eye depends on the size of the pupil, which is small in the daytime and large at night. It also depends on whether the object is a point, or a line.

In daytime, looking at a point, the resolution of the human eye can be 1–2 arcminutes — good enough to see a coin at 20 metres. From orbit, with a resolution of 2 arcminutes, an astronaut could see something as small as 150 metres across. (So the Pyramid of Khufu, which was built around 2500 years ago, and which is 230 metres square, would have been the first object built by humans that would be visible from low Earth orbit.)

However, if an object has a high contrast when compared with the objects around it, or is very straight, the human eye can see less than 1 arcminute — say 0.5 arcminute (or 30 arcseconds).

The Great Wall of China is about 8–10 metres wide. To see the Great Wall of China from the Moon with the naked eye, the resolution would have to be 0.004 arcseconds. Even the Hubble Space Telescope can only detect an object 100 metres across at the distance of the Moon.

To the naked eye, the Great Wall of China would be just visible at an altitude of about 30 km — or about four times the height of Mount Everest. If you added in the effect of a late afternoon shadow from the setting Sun, you might be able to see the Great Wall from a height of 90 km.

from the north, so the Great Wall was a military necessity. Not only did the Ming emperors repair the entire length of the Great Wall, they also added new loops. But this time they used bricks and stones. They added some 20 000 towers (mini-fortresses) and 10 000 signal beacons. A Chinese general in the Cheng Hua period (1465–1488) reported that he did not have enough manpower. To guard 500 km he had 25 camps, but each contained only 100–200 men, and one man could not guard 200 metres of frontier day and night. So Cheng Hua cleverly offered the soldiers large grants of land near the Wall, so they would be happy to be permanently garrisoned there.

After the Ming Dynasty, the Manchus allowed the Wall to fall into disrepair. The Emperor Kang Hsi ordered a Jesuit priest, Father Regis, to map his empire. It took him from 1708 to 1716. Part of Father Regis' task was to travel along the entire length of the Wall. He said that in the Chihli Province, there was only a facing of brick left. "*Along the northern border of Shansi, the Wall is made of clay without battlements, and is only about 5 feet high. West of Shansi it is a narrow mud rampart, sometimes even only a sand ridge.*"

IS THERE REALLY A GREAT WALL?

Most Westerners visit just one specific part of the Wall, at Pa-ta-ling, about 70 km northwest of Beijing. It was built by the Ming emperors. It was rebuilt in the 1950s, and today it's in absolutely superb condition. This section of the Wall is a magnificent structure built of

RADAR FROM SPACE SEES WALL

In some places, the Great Wall is so eroded and buried by centuries of blowing sand that it's not even recognisable on the ground. NASA, via their Jet Propulsion Laboratory, have been using Spaceborne Imaging Radar C/X-band Synthetic Aperture Radar (SIR-C/X-SAR) to look at the Wall from space. This radar is sensitive to vertical smooth structures such as walls — so using radar to look at archaeological structures is very effective.

Dr Guo Huadong is part of the SIR-C/X-SAR team. He is based at the Institute of Remote Sensing Applications at the Chinese Academy of Sciences in Beijing. According to him, "*Part of [the Wall] is buried by the strong winds that blow sand dunes across this part of the desert. In this region, the Wall was made out of loose soil and mud, not bricks or rocks. Usually, you cannot find these segments even if you go there, so the radar data are helping us to show the whole Wall.*"

stone and rock, about 10 metres wide and 10 metres high, even higher at the watch towers.

In fact, the Wall has never been surveyed. The path that you see on maps and atlases is just what was drawn in previous documents. The current estimate for the length of the main section of the Great Wall, from Shanhaiguan on the coast in the east to Yumenguan in the west, is about 3500 km. It's estimated that there are another 2800 km of loops and extensions in addition to this.

THE GREAT WALL ISN'T . . .

The reality is that the Great Wall of China is definitely not a huge and continuous fortification. There was never a single coherent master plan. Many Walls have been built over the last 2500 years and have then decayed.

We do have personal anecdotal evidence from various British athletes who have run along long stretches of the Wall to raise money for charity. In 1988, Edward Ley-Wilson and David Wightman ran some 1900 km of the Great Wall between Shanhaiguan and Jiayuguan Pass. Over the 47 days of their run, they averaged about 41 km per day. A Mr H.J.P. Arnold wrote to Mr Ley-Wilson inquiring about the integrity of the Great Wall, and got this reply:

"*Wall clearly discernible and only moderately eroded or broken along its length — 22% of the 1,900 kilometre run.*

Wall usually discernible but frequently broken/eroded 41%.

Wall scarcely discernible, almost totally eroded and running by reference to maps 37%."

In fact Mr Arnold wrote to dozens of astronauts and asked them if they had been able to see the Great Wall from the Moon or low Earth orbit. Mr Arnold has done a magnificent analysis of whether we can see the Great Wall of China from space. It was published in the magazine *Spaceflight*.

The Chinese have been building major walls for at least 2500 years. Over this time, various walls have been allowed to fall into disrepair ... only to be rebuilt later.

Today, some sections of the Wall are in as-new condition, while others cannot even be recognised as having once been a wall.

ASTRONAUTS: GREAT WALL FROM THE MOON

Neil Armstrong, commander of Apollo 11, said about the Great Wall of China, "*It is not visible from lunar distance.*" Edwin "Buzz" Aldrin, who piloted the Lunar Module, said, "*You have a hard time even seeing continents.*"

James Lovell, who flew on Apollo 8 and was the commander of the ill-fated Apollo 13, said that to "*spot the Great Wall ... with the naked eye from the vicinity of the Moon is absurd!*"

THINGS YOU CAN SEE FROM SPACE

From low Earth orbit, you can see motorways (especially at night when they're illuminated), various large dams and circular agricultural projects. Andy Thomas wrote of seeing "*... the fencing off of farm land into individual fields ... in the Midwest of the US and Canada. One of the most visible signs of human presence is the occurrence of contrails (cotton trails) from aircraft in the upper atmosphere ... seen over virtually all parts of the world as white streaks across the sky.*"

Alan Bean, the Lunar Module pilot for Apollo 12, said, "*The Great Wall of China is most definitely not visible to the unaided human eye at lunar distances.*" He is further quoted in Tom Burnam's book *More Misinformation:* "*The only thing you can see from the Moon is a beautiful sphere, mostly white (clouds), some blue (ocean), patches of yellow (deserts), and every once in a while some green vegetation. No man-made object is visible on this scale.*"

And Harrison Schmidtt, the Lunar Module pilot for Apollo 17, wrote: "*Only desert areas and desert or non-green coastlines are clearly defined to the eye from the Moon.*"

In 1988, Edward Ley-Wilson and David Wightman ran about 1900 km of the Great Wall. They reported that for 37% of the distance run, the Wall was scarcely discernible and was almost totally eroded. In fact, they often ran by map references.

THE GREAT WALL FROM LOW EARTH ORBIT

So we can't see the Great Wall from the Moon, which is about 400 000 km away. But could you see it from the Space Shuttle? It flies in low Earth orbit, 300–530 km up.

The astronaut William Pogue, who flew in space on Skylab 4, was able to see the Great Wall, but only with binoculars and with lots of practice. He wrote a book called *How Do You Go to the Bathroom in Space?* He writes, "*It wasn't visible to the unaided eye. The first time I thought I had seen it, I was in error. It was the Grand Canal near Peking. Later, I was able to identify the faint line of the wall, which zig-zags in a peculiar fashion across hundreds of miles.*"

The American astronaut Sally Ride talked about how the high speed of the Space Shuttle makes it very difficult to see anything down on the ground. "*I found that at this speed (5 miles per second), unless I kept my nose pressed to the window, it was almost impossible to keep track of where we were at any given moment — the world below simply changes too fast.*"

Karl Henize flew on Space Shuttle 51F in July 1985 — and he too failed to see the Great Wall from orbit. Alta Walker from the US Geological Survey wanted him to look for the Wall, and so helped him out with notes and maps. The Space Shuttle is so fast that Karl Henize had only a very short time to look for the Wall. Afterwards, he wrote: "*There is little time for map reading — from the time an approaching area is 45° from the [vertical] — at which time one can begin to make out a fair amount of detail — until it is slipping under the spacecraft hull is 40 seconds at most.*"

Even Neil Armstrong, commander of Apollo 11, said that the Wall was not visible from the Moon. In fact, you have a hard time even identifying continents.

The astronaut Andy Thomas wrote about recognising objects on Earth on May 22, 1998, while he spent 141 continuous days on the Russian space station Mir. He wrote that what was striking was "*the abundance of clouds*". After a while, he could recognise the deserts of northern Africa, the redness of Australia and characteristic coastlines such as the "boot" of Italy. But even though he tried, he was never able to see the Great Wall. He wrote: "*Cities can be seen, although, surprisingly, they do not stand out readily. But we can make out their grid-like pattern of streets. The stories about the Great Wall of China being visible from space may be true, but I have yet to see it.*"

The astronaut James Lovell said that, with the naked eye, he could recognise El Paso airport from orbit, because he had flown over it many times in various aircraft. But he could not recognise the similar-sized airport at Kinshasa in Zaire even though he knew where to look — because he was not familiar with it.

So maybe Chinese astronauts will be the first to regularly see the Great Wall of China with the naked eye …

DID THE GREAT WALL WORK?

No. For all its reputation, the Great Wall never stopped the invaders. It was only effective against small, petty raids. But it never stopped the huge movements of the Tartar hordes.

References

H.J.P. Arnold, "The Great Wall: Is it or isn't it?", *Astronomy Now*, April 1995, p 10.

H.J.P. Arnold, "The Great Wall of China from space — The exploration of a myth", *Spaceflight*, Vol. 31, July 1989, pp 248–252.

"Chinese Puzzle", *The Last Word*, New Scientist, London, 1998, pp 82–84.

Andy Thomas, "The view from space", *Letters From the Outpost*, 22 May 1998 — spaceflight.nasa.gov/history/shuttle-mir/ops/crew/letters/thomas/letter5.

Adam Warwick, "A thousand miles along the Great Wall of China — The mightiest barrier ever built by man has stood guard over the land of China for twenty centuries", *National Geographic*, February 1923, pp 113–143.

The Loch Ness MONSTER

Does Nessie really exist?

In these scientific days, people are pretty sceptical about monsters. But Nessie, the Loch Ness monster, has always had a lot of credibility. First, tourists and locals kept on seeing something strange surging to the surface of Loch Ness. Second, there was that famous fuzzy photo showing its long neck rising out of the waters of Loch Ness. Everybody thought that the photo was genuine, because it was taken by a Pillar of the Community, the ex-military London medical doctor Colonel Robert Kenneth Wilson.

What most Nessie-believers still don't realise is that the famous never-since-duplicated first photo was a fake.

LOCH NESS — HISTORY

Loch Ness is actually part of a major geological fault line that cuts through Scotland. The Loch "belongs" to two very different land masses. The southeast side of the Loch is definitely in Europe. But the land on the northwest side of the Loch was left behind from North America, when the Atlantic Ocean opened up several hundred million years ago.

Quite recently, about 10 000 years ago, a glacier pushed through the Loch. That makes it highly unlikely that Nessie is a plesiosaur left over from the days of the dinosaurs. Fishy creatures don't usually survive well in solid ice.

Furthermore, the dinosaurs and their relatives died out about 65 million years ago — 6500 times further back than the glacier.

The legend of the Loch Ness monster goes way back. Back in the Middle Ages, there was a tale that the Loch contained a mysterious creature called a "water horse" or a "kelpie", which would supposedly lure travellers to their death.

The very first recorded "sighting" was in the year AD 565. St Columba chanced upon the funeral of a man who had swum in the Loch — and had been bitten to death by this Loch Ness monster. Another report says that some time later St Columba actually saw the Loch Ness monster with his own eyes. According to the tale, he boldly told the "fearsome beastie" to behave itself. His incantations appear to have worked, as there have been no more reports of the monster attacking people.

NESSIE — THE PHOTO

Loch Ness remained relatively isolated for some 1400 years, but in early 1933 a road was built past the Loch. In April 1933 Mr and Mrs Mackay drove along this new road by the side of the Loch. They were astonished to see the water surging and boiling in the centre of the Loch for several minutes. The *Inverness Courier* reported that they saw "*an enormous animal rolling and plunging*". The news of this modern viewing of "Nessie" rapidly spread around the world.

Shortly after, in 1934, the respected Harley Street gynaecologist Colonel Robert Kenneth Wilson released a fuzzy black-and-white photo. He explained that he had been driving on the new road around Loch Ness on the morning of 19 April. To his surprise, the water near him began to surge and boil. The *Daily Mail* newspaper reported how he quickly grabbed his camera and took two photos. Neither photo was very sharp. One photo showed Nessie's head slipping beneath the waves, but the other one showed a long neck and head stretching up out of the waters of Loch Ness.

That photo, fuzzy as it was, gave the legend of the Loch Ness monster an enormous boost. The evidence for "Nessie" seemed to be very solid.

NESSIE — THE FILM

At 9 am on 23 April 1960, an aeronautical engineer, Tim Dinsdale, was driving on the road around the Loch. He saw what seemed to be an object with a hump about 1.2 km away. He took four minutes of black-and-white film footage. As he shot his footage, the object seemed to swim away from him towards the opposite shore, leaving behind a definite V-shaped wake. Then it suddenly vanished. Dinsdale had the film analysed at the RAF's Joint Air Reconnaissance Intelligence Centre.

NESSIE — THE CONFESSION

But the legend took a beating in November 1993, when a certain Mr Christian Spurling confessed on his deathbed that he had made the monster with his own hands. He told David Martin, a former zoologist with the Loch Ness and Morar Scientific Project, and his fellow researcher, Alastair Boyd.

The scam had been set up by

Mr Marmaduke "Duke" Wetherell. He was a film-maker, actor and self-styled "big-game hunter". Mr Wetherell had been hired by the *Daily Mail* newspaper to find this headline-making monster. But instead of looking for the monster, he decided it would be easier to make one. So he sent a message to his stepson, Mr Christian Spurling, asking, "*Can you make me a monster?*" Christian Spurling was a professional model-maker. He later said, "*I just sat down and made it. It was modelled on the idea of a sea serpent.*"

Christian started off with a toy tin submarine, costing only a few shillings at the local Woolworths store in the London suburb of Richmond. He shaped some plastic wood to make a long neck and a small head. The finished monster was about 45 cm long and about 30 cm high. It had a lead keel to keep it stable in the water. They went down to the water on a quiet day and floated the model out into the shallows. Just to keep the whole scam in the family, Mr Wetherell's other son, Ian, took the photo.

The brilliant finishing touch was to give it an air of respectability by getting a London surgeon, Robert Kenneth Wilson, to be the front man. But once the photo got into the newspapers, their little joke got completely out of hand — and so the myth of Nessie was rekindled.

NESSIE — THE DEBUNKING

Dr Maurice Burton wrote a series of three articles on Nessie in the *New Scientist* magazine in 1982. He contacted Wilson, whom the newspapers said took the famous photograph. Suddenly he heard some serious back-pedalling. "*When I contacted Wilson himself, in quest of further details, he replied tersely that he made no claim to having photographed a monster, and did not believe in it anyway.*"

But if that Nessie photo was a fake, how do you explain all those sightings — on average, one every 130 hours of observing?

THE CALEDONIAN CANAL

In 1773, the British Government employed James Watt to make a series of surveys for canals to link the North Sea (via Moray Firth) with the North Atlantic (via Loch Linnhe). The proposed canals would join various lochs, including Loch Ness.

Thomas Telford directed the start of construction in 1803. The system started carrying traffic in 1822, and was completed in 1847. The total length of the "Caledonian Canal" today is about 100 km, made up from some 37 km of constructed canals, with the remainder being the pre-existing lochs.

The Caledonian Canal used to have a major economic importance, but it is too small for today's ocean-going ships. Today, the Caledonian Canal is used by tourist, fishing and pleasure craft.

Oooot ... it ain't Nessie!

It's actually all about the trees ... the Scots pine, **Pinus sylvestris** to be exact.

The Scots pine is a petrochemical plant. To protect it from the harsh Highland conditions, the tree makes tar oils, turpentine and lots of resins.

The pine falls into the Loch, then becomes "waterlogged" and eventually sinks.

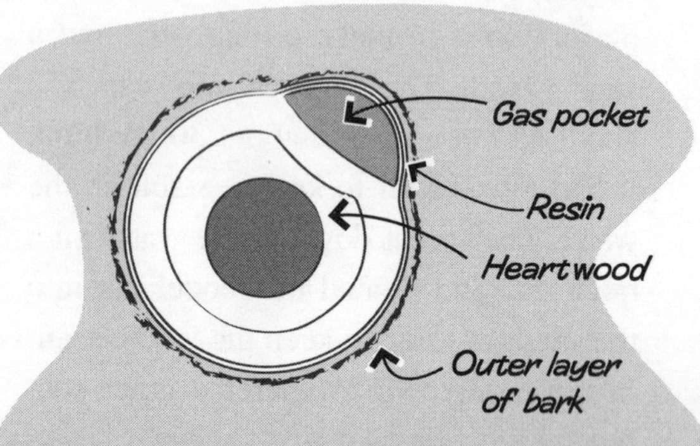

At the bottom of the Loch, gas forms as the log decays and the gas bubbles force the resin out of the bulk of the log, making it look somewhat monster-like.

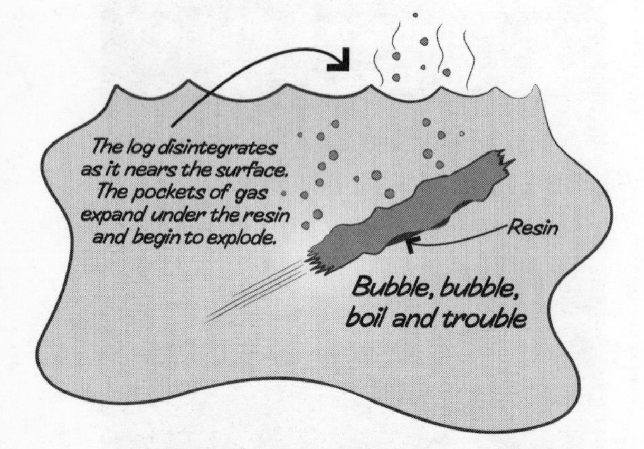

Because the end of the log isn't sealed with resin and bark, the gas escapes, thereby propelling the log both upwards and across.

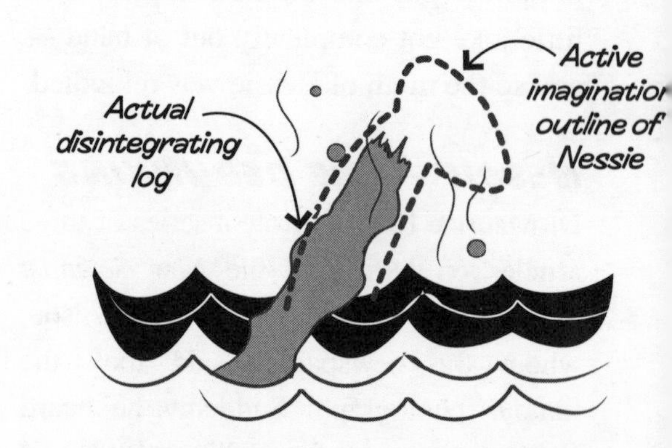

An overexcited mind can mistake all this activity for the famous "Nessie" emerging to say hello to the bonny folk on the shore!

THEORY NO. 1

One popular theory is that when a family of otters frolic in the water, distant viewers on the shore see their backs as the humps of a giant monster. This theory would explain many of the sightings. Otters can be up to 2 metres long, and they often travel in families. A family of otters swimming in line would appear as a series of humps in the water. Indeed, a typical description of a swimming otter family is "*like an upturned boat towing three upturned dinghies*". And if the leading otter lifts itself up to get a better view, then presto! You "see" the head and neck of a long sea monster ...

Otters can disturb the water mightily. Burton writes in the *New Scientist* of how the otter "*suddenly comes up to the surface, and proceeds to writhe, somersault and roll within a limited area, as if in ecstasy, throwing water in all directions in a flurry of foam and spray*".

THEORY NO. 2

Robert P. Craig, a Scottish electronics engineer, has a second, and much more interesting, theory, which he published in the *New Scientist* back in 1982. He ignored otters, and instead pointed the finger at the Scots pine *(Pinus sylvestris)*.

Here is an important clue. There are more than 500 lochs in Scotland, but only three of them are associated with monsters: Loch Ness, Loch Tay and Loch Morar. These lochs have two characteristics in common: they are all very deep, and they were all once surrounded by large forests of Scots pine. Today, there are only small remnants of Scots pine around these three lochs.

So Robert Craig's theory starts off a century or more ago, with a 1-tonne Scots pine tree falling into the lake, becoming "waterlogged" and eventually sinking. At the bottom of Loch Ness, 250 metres down, the pressure is about 25 atmospheres. (Twenty-five atmospheres of pressure, in the boiler of a steam locomotive, will push it faster than 160 kph.) And in Robert Craig's theory, this pressure is the source of the energy that causes the sightings.

Down at the bottom of the Loch, our Scots pine log is gradually decaying. Tiny bubbles of gas are forming. The Scots pine is a tree — but it's also an organic petrochemical plant. To shrug off the harsh Highlands climate, this tree makes tar oils, turpentine and lots of resins. These resins trap the bubbles and stop them from floating to the surface. In fact, the bubbles can force the resin out of the bulk of the log. The resin is sticky, and remains stuck to the log. So the log takes on a grotesque shape as it gradually grows excrescences of resin that are full of bubbles.

FRESH WATER

Loch Ness is the largest freshwater lake in the United Kingdom. It is about 36 km long and 2 km wide, reaches depths of 240 metres, and carries about 7.4 cubic km of water. Loch Ness is in the Inverness District in the Highlands of Scotland.

THE POWER OF PRESSURE

The pressure inside these bubbles is the same as on the outside (25 atmospheres), so they don't get the chance to grow very large. Each of these little bubbles is a tiny buoyancy tank. Eventually, there are enough of these gas bubbles to lift the log off the bottom of the Loch.

Moving slowly at first, the log begins its trip to the surface. The log looks like a log for most of the trip. But in that last 50 or so metres it begins to disintegrate. That's because the pressure is still 25 atmospheres on the inside of the log, but only 5 atmospheres on the outside. So the bubbles swell enormously. At some depth in that last 50 metres, the outward force of the bubbles is greater than the structural integrity of the log and it starts to burst at the seams. The bubbles literally explode the log into a bunch of splinters — some large and some small.

And so a great, frothing, boiling mass bursts through the surface of Loch Ness. This explains the terrific upsurge of water that several observers have been lucky enough to see. Within a few moments, the waterlogged pieces of wood head back down to their watery grave, while the bubbles of gas simply vanish into the atmosphere. And the head or flippers that some people see? Well, they're just odd-shaped lumps of wood (if they are not otters). And the resin would take on many strange shapes under the influence of the bubbles.

There's no bark sealing the end of the log, so the gases can escape from there more easily. It would be an underwater rocket, with gases coming out in one direction — and the log being pushed in the other direction. As the log surged to the surface, it would have a horizontal velocity as well as a vertical velocity. So the bubbling, boiling, frothing mass would keep that horizontal velocity. Instead of just coming to the surface and staying there, it could move across the water before sinking again.

So the Loch Ness monster might be just a log Ness monster . . .

PRESSURE

Normal atmospheric pressure pushes on a 1-metre-square window with 10 tonnes of weight. The reason why the window stays intact is that there's another 10 tonnes of weight on the other side of the window pushing back.

Another way to look at it is to think of a square, 1 metre on each side, in your back yard. The total weight of all the air directly above that square, from the ground up to the edge of space, is 10 tonnes.

The pressure in your car's tyres is about 2 atmospheres, or 20 tonnes of force pushing on each square metre.

In water, the pressure increases by 1 atmosphere for every 10 metres that you dive. At the bottom of Loch Ness, the pressure is about 25 atmospheres — roughly equal to 250 tonnes of force pushing on each square metre.

THEORY NO. 3

But what about the black-and-white film footage that showed something leaving behind a V-shaped wake? The RAF's Joint Air Reconnaissance Intelligence Centre advised that the film be submitted to experts in hydrodynamics, who analysed stills from the film. They said that it appeared to be "*a small body moving over the water*".

In 1960, when Dinsdale shot his footage, motorboats would often cross the Loch early in the morning. These motorboats were similar in many ways to Dinsdale's "monster". They had the same speed, size and shading, and left the same wake and wash. Dr Maurice Burton wrote in his third *New Scientist* article that the creature "*was said by Dinsdale to have submerged at the spot where the boats I watched crossing over, in 1960, shut off their motor, turned hard towards the beach and disappeared suddenly under the overhanging branches of trees*".

HAIR TURNS WHITE . . .

The *National Geographic* reports that in "*the 1930's, a team of divers brushed against something huge lurking on the bottom and surfaced with their hair turned snowy white*" (see *Hair Turns White Overnight*).

NOT ENOUGH FOOD

Loch Ness doesn't have enough food to support Nessie.

The Loch doesn't get a lot of sunlight, because it's at a high latitude and is often clouded over. The water is murky with suspended peat, so any sunlight that does hit the surface penetrates only 6 metres or so. The steep hills around the Loch don't yield many nutrients to the fast-flowing streams that feed the Loch.

The combination of all these factors means that there aren't many plants for the small fish to feed on. Two separate estimates of the total mass of the fish in Loch Ness are quite similar: 20 tonnes and 30 tonnes.

If Nessie were the size that is claimed, she would weigh at least 1 tonne and would need to eat about 100 kg of fish each day. A Nessie would not pop into existence by itself — it would need a mother and father. The smallest estimate of a viable population size is about 10.

So 10 Nessies would eat about 1 tonne of fish per day, emptying Loch Ness of food in roughly one month . . .

THE FULL ANSWER

But why, then, have so many people gone looking for Nessie? Perhaps the search for Nessie has always been a question for the psychologists, not the zoologists.

In today's enlightened times, we have looked back in time to shortly after the Big Bang, and we have mapped the human DNA. Even so, we still hunger after myth, magic and mystery — and Nessie gives us all of these. And the tourists do spend about US$40 million each year in and around Loch Ness.

SEA MONSTERS

We humans have long believed in enormous snake-like marine creatures.

In the Bible, the Old Testament talks about battles between God and Leviathan (also called Rahab). A Canaanite poem from Ugarit (which today is Ras Shamra in northern Syria) also records a battle between the god Baal and a sea monster called Leviathan. Babylonian literature says that the god Marduk had a battle with Tiamat, a multi-headed serpent-dragon. A Hittite myth records how the weather god beat the dragon Illuyankas.

References

Dr Maurice Burton, "The Loch Ness saga. A fast moving, agile beastie", *New Scientist*, No. 1312, 1 July 1982, pp 41–42.

Dr Maurice Burton, "The Loch Ness saga. A flurry of foam and spray", *New Scientist*, No. 1313, 8 July 1982, pp 112–113.

Dr Maurice Burton, "The Loch Ness saga. A ring of bright water", *New Scientist*, No. 1311, 24 June 1982, p 872.

Robert P. Craig, "Loch Ness: The monster unveiled", *New Scientist*, No. 1317, 5 August 1982, pp 352–357.

William S. Ellis and David Doubilet, "Loch Ness. The lake and the legend", *National Geographic*, June 1977, pp 759–778.

"Nessie never posed", *Time*, 28 March 1994, p 19.

Adrian Shine, "The biology of Loch Ness", *New Scientist*, No. 1345, 17 February 1983, pp 462–467.

Hair Turns WHITE OVERNIGHT

Is there a chance that my hair might turn white overnight?

Thomas More was honoured by King Henry VIII for his services. But Thomas More later refused to acknowledge that King Henry VIII was the Head of the Church of England. He was convicted of treason and beheaded on 6 July 1535. It is said that his hair turned white on the night before the execution. Is this possible?

Writers think so. They have written about sudden traumatic events turning your hair white overnight for centuries.

HISTORY OF WHITE OVERNIGHT

Sir Walter Scott writes in *Marmion*:

For deadly fear can time outgo,
And blanch at once the hair.

In Shakespeare's *King Henry IV*, Sir John Falstaff says to Hotspur:

Worcester is stolen away tonight; thy father's
Beard is turned white with the news.

Byron writes in his poem *The Prisoner of Chillon*:

My hair is grey, but not with years,
Nor grew it white
In a single night,
As men's have grown from sudden fears.

The red hair of Mary, Queen of Scots, is also supposed to have turned white just before her execution. Wordsworth writes in his *Lament of Mary, Queen of Scots*:

Those shocks of passion can prepare
That kill the bloom before its time;
And blanch, without the owner's crime,
The most resplendent hair.

That distinguished look!

For many, finding a white hair can be a distressing event. BUT imagine your hair turning white OVERNIGHT!

The concept of hair-turning-white-overnight is also in common language. In English, we have the saying "*It's enough to make your hair turn white.*" The German language has an equivalent, "*Sich graue Haare etwas wachsen lassen.*"

In India, Mumtaz Mahal was the favourite wife of Shah Jahan. She died in his arms on June 17, 1631, after bearing him 14 children. His hair supposedly turned white within a fortnight. To honour her memory, he built one of the most beautiful buildings in the world, the Taj Mahal.

John Libeny tried unsuccessfully to assassinate the Emperor Franz Josef of Austria. The *London Times* wrote: "*His hair, which was previously black, had become nearly snow-white in 48 hours preceding his execution; it hung wildly about his head; his eyes seemed to be starting from their sockets and his whole frame was convulsed.*"

So the literati like to think that fear can whiten more than just your complexion.

THE VOICE OF REASON

But how can this be? Hair is dead tissue. The hair follicle pushes it out at about 1 cm per month. The dark colour comes from a chemical called melanin, which is added in the hair follicle in the scalp, a long way from the tip of the hair. Hair is "dead", and the only way our technology can make it change colour is with dyes.

So how do 120 000 black hairs each turn white overnight, from base to tip?

The answer is easy. They don't. Instead, first, as a result of natural ageing, some of the dark hairs have already turned white. Second, some sudden event makes the dark hairs fall out, leaving the white hairs behind.

A HAIR-RAISING TECHNIQUE TO CLEAN UP OIL

Tonnes of human hair get cut every day, and end up in landfill. Phillip McCrory, a hairdresser from Madison, Alabama, discovered one good use for these lost locks.

He was watching TV footage of 1989's big oil spill in Prince William Sound, Alaska. *"I saw an otter being rescued whose fur was saturated with oil. I thought, if animal fur can trap and hold spilled oil, why can't human hair?"*

A lesser human would have left the idea as just an idea, but McCrory went one step further and *Did the Experiment*. He stuffed about 2.3 kg of his customers' hair into his wife's pantyhose, and tied them into a doughnut shape. He then filled his son's wading pool with water, floated the "doughnut" of hair-filled pantyhose on the water, and tipped dirty used engine oil into the centre of the doughnut.

He pulled the circle shut. As it closed, it soaked up the oil. He lifted out the hair-filled pantyhose, and saw that there was no oil left in the water. The hair had "adsorbed" the oil. "Adsorb" means to absorb, but only on the surface. He could remove the oil by squeezing the hair, and then use it again. He claims that 567 grams of hair can soak up 3.78 litres of oil from water in less than two minutes. Current adsorbers, such as polypropylene, take between one and two days to soak up that much oil.

Duck feathers and sheep wool also trap oil, but they are expensive and don't adsorb as well as human hair. McCrory tested his concept with the nearby Marshall NASA Center Technology Transfer Office. He claims that his technique can drop the cost of cleaning up oil spills from US$10 per gallon to US$2. He has now set up his own company, BEPS Inc, to market his invention.

HAIR CYCLE

A hair is a long, thin fibre made of the protein keratin. You find keratin in skin, nails, horns and hooves, and even some hair shampoos. Each hair is made by a single hair follicle.

Hair follicles continually cycle through three stages.

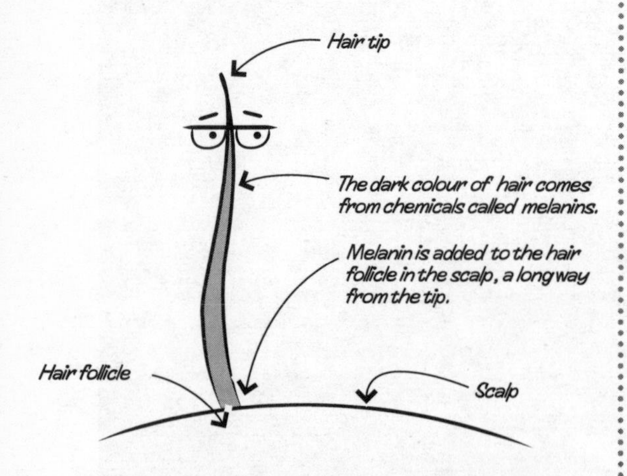

Hair is dead tissue. The hair follicle pushes it out at about 1 cm per month.

The first stage is the long growing period (anagen). At any given time, about 80–90% of the hairs are in the anagen stage. The time that a hair follicle stays in anagen depends on the person, their age (eg, infant or hormone-laden teenager) and the hair's location (eg, eyebrow or scalp). A hair follicle on your eyelid will stay in the growth phase of anagen for a short time only — which is why eyelashes are usually short.

A scalp follicle can grow hair for six years in women, but only four years in men. In that time, a single hair can grow 75 cm in a woman, but only 45 cm in a man. That's why women can have longer scalp hair than men. At the beginning of pregnancy the hormone levels change enormously, and so the number of hairs in anagen drops.

The second stage (catagen) is where the hair withdraws and separates from the matrix (part of the hair-making

ALOPECIA AREATA

Alopecia areata is an enigmatic, poorly understood, autoimmune, non-scarring disease of hair loss. It commonly appears in sharply-demarcated circular areas about 2–5 cm across, which will usually be completely devoid of hair. In severe cases, these circles can join up. The condition affects about 0.15% of the population. Unlike most other autoimmune diseases, there is usually no permanent tissue damage.

Alopecia areata is sometimes associated with hyperthyroidism, hypothyroidism, diabetes and Down's Syndrome. In terms of age at first appearance, there are two peaks: one at the age of five, the second at 30 years of age. In mild cases the hair loss might not even be noticed, especially if it doesn't happen on the scalp.

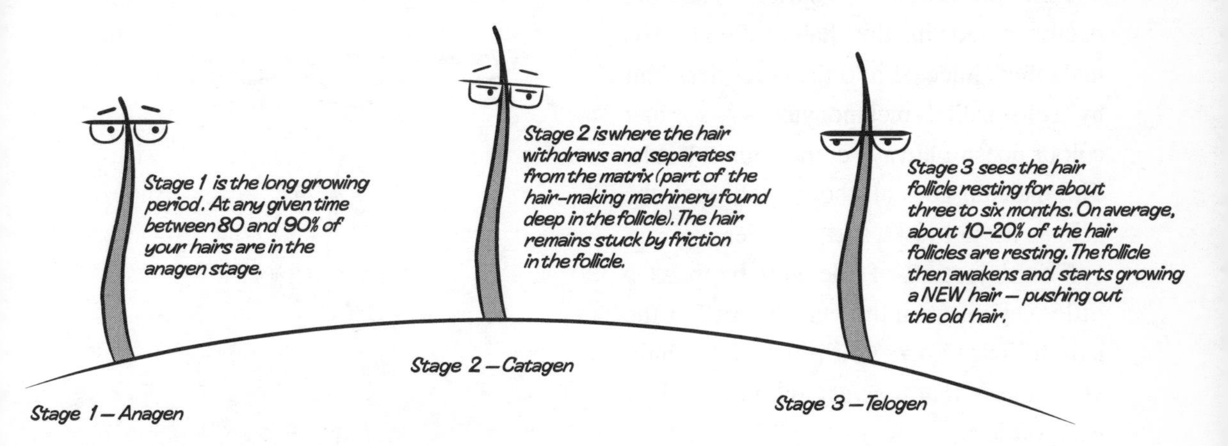

Now, a hair is a long, thin fibre made of the protein keratin. Keratin is found in skin, nails, horns, hooves and even some shampoos. Each hair is made by a single hair follicle. AND ... a hair follicle cycles repeatedly through three stages.

machinery, deep in the follicle). The hair remains stuck by friction in the follicle — but a tug will remove it. Catagen lasts about two to four weeks. At any given time, fewer than 1% of the hairs on your scalp are in catagen.

In the third stage (telogen), the follicle has a rest for about three to six months. During pregnancy, the number of hairs in telogen increases. On average, about 10–20% of the hair follicles are resting.

Then the hair follicle wakes up and starts growing a new hair — and this pushes out the old hair. You lose about 100 hairs this way each day.

COLOUR OF HAIR

As a general rule, everything in the body is more complicated than you think. So there are many different chemicals that are called "melanin", such as "eumelanin" and "phaeomelanin". Melanin is as vague a chemical term as "wood".

9.6-TONNE HAIR

We humans are just hairless apes — but we do spend a lot of time thinking about our hair.

We have about 120 000 individual hairs on our scalp. Each hair is about 70 microns in diameter (a micron is a millionth of a metre) and can support an 80-gram weight.

If you could weave all of your scalp hairs into a single rope, it could carry a 9.6-tonne truck. So circus performers who hang by their hair are working with a safety factor of about 140 (ie, their hair is able to support 140 times their own weight).

Hairs are actually colourless. They are manufactured in the hair follicle. The melanin is injected into the colourless hair by cells called melanocytes — so hair colour doesn't always come from a bottle. The melanin goes all the way through the hair — it's not just a surface layer.

This injection of melanin happens a little way up from the very bottom of the hair follicle. So if you pluck out a hair, you can see that the very base of the hair is colourless.

Very dark black hair has lots of eumelanin. Blond or red hair has phaeomelanin. Colours between black and blond happen because of varying amounts of both melanins. The hair of albino people has no melanin at all.

There is no such thing as an individual grey hair ... what looks like grey hair is a mixture of dark hairs and white hairs.

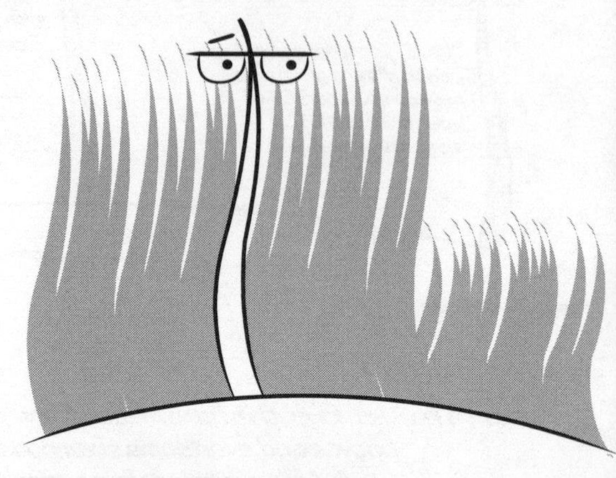

White hair comes about due to a complete loss of melanin. These hairs take on a translucent, whitish colour.

GREY HAIR

What happens with "grey" hair is that you have a complete loss of melanin in some individual hairs. It is quite common in Westerners, but quite uncommon in South American Indians. We don't know why. Those hairs take on a translucent white colour. But the rest of the follicles of the scalp keep making dark hairs. Sometimes grey hair is called "salt-and-pepper" hair.

So here is a Great Truth — there is no such thing as individual grey hairs. Grey hair is an optical illusion. It's actually a mixture of ordinary dark hairs with melanin, and those special translucent hairs without melanin. But let's just call it "grey" for convenience's sake.

Grey hair usually begins on the temples at the side of the head, spreads to the scalp, and then finally to the beard and moustache (if you're a man) and eyebrows.

COMMON BALDNESS

Common baldness is a natural, slow, painless loss of hair that increases with age. About 60% of men have significant hair loss by the age of 50, while 50% of women have some hair loss by the age of 60.

In male-pattern baldness, the hairline at the forehead and temples recedes first, then the hair thins over the crown (at the top, and back, of the skull). Women don't generally get recession of the hairline at the front. Instead, they suffer a generalised thinning on top.

The name "temple" comes from the Latin for "time" — "tempus". You usually see those first grey hairs (the early marks of time) on a person's temples.

Normally grey hair is inherited, and appears at any time between 15 and 50 years of age, as the production of melanin slows. About 25% of people have some grey hairs by the time they are 25 years old. The amount of grey hair increases as you get older. Grey hair can also be caused by thyroid disease, pernicious anaemia and diabetes.

There are a few conditions that can make the dark hairs fall out. We don't know why they do this and we don't know why the white hairs stay behind — the answers might win somebody a Nobel Prize.

TELOGEN EFFLUVIUM

The most common cause is called "telogen effluvium". Some kind of stress somehow makes the hair follicles shift from the growing phase (anagen) to the resting phase (telogen).

There are very many different types of stress, besides loss of a loved one or a sudden change in lifestyle. Other stressors include changes in hormone levels (post-pregnancy, hyper- or hypothyroidism), nutritional abnormalities (deprivation of calories or protein, or deficiencies in biotin, essential fatty acids, iron or zinc), or drugs (anti-coagulants, beta blockers, lithium and even too much Vitamin A). They also include physical stressors, such as surgery, or a sudden generalised disease.

TRANSPLANT HAIR FOLLICLES

It seems as though hair follicles can be transplanted from one person to another.

Back in 1999, Colin Jahoda and Angela Christiano from Durham University in the UK *Did the Experiment*. They removed a few hundred hair follicle dermal-sheath cells from the base of a hair follicle in Colin's scalp, and transplanted them into Angela's forearm.

Amazingly, her immune system did not reject these cells from a man. And after about 24 days, new hair follicles began to grow, which actually sprouted new hairs. These hairs were larger and thicker than Angela's original forearm hairs, mostly pigmented, and grew in several different directions. The scientists later analysed cells from the base of her new hair follicles, and found that the cells had both X and Y chromosomes — in other words, the cells came from a male.

As a result of this experiment we might have a new treatment for hair loss, and perhaps a deeper understanding of our immune system.

We don't know why the hair suddenly falls out. One theory is that in times of stress the body likes to conserve its resources, so the hair follicles stop putting energy into growing. Shortly afterwards (usually four to six weeks, but sometimes only a week or less), when the stress is over, the hair follicle starts regrowing — and the new hair pushes out the old one.

You can also see this stop-start effect in the fingernails. Stress (such as sudden short illness) can stop the fingernails from growing. A few weeks later, you can see evidence of this stress in a horizontal ridge on the nails called the Beau's Line.

SURE-FIRE CURE FOR BALDNESS

There is a drastic cure for male baldness.

In males, baldness is programmed by the male sex hormone, testosterone. Without testosterone, a male can't go bald. So if a male is castrated at an early age (before puberty), he never makes any significant quantities of testosterone. And the lucky eunuch can carry a full head of hair throughout his entire life.

ALOPECIA AREATA

A less common cause of hair loss is the disease called "alopecia areata". Here your hair, for no apparent reason, simply falls out. Sometimes, this is genetically inherited. Alopecia areata can happen after a major life crisis, but in most cases it just happens with no known cause.

In some cases of alopecia areata the dark hairs fall out — giving you "instant" white hair over a few days or weeks.

MAYBE CURE FOR BALDNESS

An important thing to realise is that both men and women have both male and female sex hormones — but men have more male hormones, and women have more female hormones.

The enzyme 5-alpha-reductase turns testosterone into a more potent form of testosterone called "dihydrotestosterone" (DHT). DHT acts on receptors on some of your cells. In general, in women, the DHT receptors are less active than the DHT receptors in men.

If your hair follicles happen to be sensitive to DHT they will shrink, setting off baldness. There is a drug that blocks the action of the converting enzyme — so it will reduce the level of DHT and its effects.

We think that one way someone's hair can turn "white overnight" is this. Firstly, due to natural ageing, some dark hairs have already turned white. Secondly, a sudden traumatic event makes the dark hairs fall out. Bingo, "instant" white hair.

HAIRY NOSES AND EARS

As men get older, they lose hair on the top of their skull and start sprouting it from their noses and ears. Why is this so?

First, there are two *forms* of hair. Terminal hair is your regular hair — it has a central core and its hair follicles are deeply rooted. Vellus hair is fine hair, because it doesn't have the central core. Its hair follicles are quite shallow. It grows in the areas that do not normally grow a lot of hair. As you get older, some of the terminal hair follicles on your scalp turn into vellus hair follicles, before they finally close up and vanish.

Second, there are three *types* of hair. First, there is non-sexual hair, which both boys and girls have (scalp, eyebrows, legs, etc). Second, there is ambisexual hair. It occurs in both sexes, once the hormones of puberty kick in (eg, armpit and pubic hair). Third, there is male sexual hair, which is related to male sexual hormones (eg, beard and chest hair).

Both forms of hair, and all three types of hair, are different in how they respond to different hormones. As you go through childhood and puberty into adulthood, and then into old age, the ratios of your various hormones change.

And so the hairs on your scalp get thinner (terminal to vellus), but the ones in your nose get more robust (vellus to terminal). If only it were the other way around . . .

Alopecia areata is different from common balding, because it is just as prevalent in women as in men. Alopecia areata is also more common in children than in adults.

Currently, there is no successful treatment available for alopecia areata. One difficulty faced by children with the condition is that people think their baldness is due to leukaemia. So the kids get treated as though they are about to die!

THE TAKE-HOME MESSAGE

So you can (very rarely) have your hair turn white — it won't happen overnight, but it can happen . . .

HAIRDRESSERS 3500 YEARS AGO

About 3500 years ago, the Assyrians lived in what is today northern Iraq. They had great skill in tinting, oiling, perfuming, layer-cutting and even hot-rolling of hair.

Baldness was a sign of weakness, so wigs were the standard way to cover up a shiny scalp and chin. Women could appear in the Royal Court on official business only if they wore a fake beard.

References

J.E. Jeliner, M.D., "Sudden whitening of the hair", *Bulletin of the New York Academy of Medicine*, Vol. 48, No. 8, 1972, pp 1003–1013.

Dr Stephen Juan, *The Odd Body: Weird and Wonderful Mysteries of our Bodies Explained*, HarperCollins Publishers, Sydney, 1995, pp 125–126, 140–151, 153–159, 212–213.

Dr Stephen Juan, *The Odd Body 2: More Weird and Wonderful Mysteries of our Bodies Explained*, HarperCollins Publishers, Sydney, 2000, pp 42, 84–85, 124–135.

"NASA tests hair-raising technique to clean up oil spills", *NASA News*, 22 April 1998.

Amanda Reynolds et al, "Trans-gender induction of hair follicles", *Nature*, Vol. 402, 4 November 1999, pp 33–34.

Belly BUTTON BLUES

What on Earth is the stuff That makes up Belly Button Fluff?

It all began innocently with a "simple" question on my Triple J Science Talkback radio show (Thursdays, 11.00–12.00). "*Why is my Belly Button Fluff blue — and why do I get it, anyhow?*"

This was a real question, and it deserved a real answer. So we went looking. We couldn't find any surveys involving large numbers or any hard research involving microscopes. But we did find a few opinions.

EARLY THEORIES OF BELLY BUTTON FLUFF (BBF)

Research always starts off with a few tentative probes into the unknown. Later people benefit from this early work. Isaac Newtown said, "If I have been able to see far, it is because I have stood on the shoulders of giants." Tim Albert was one of these giants.

Back in August 1984, Tim Albert was the Editor of the *British Medical Association News Review*. A researcher from a BBC radio program had asked the BMA office why men (but not women) get bits of blue fluff in their navels. The *BMA News Review* posed this question to its readers, and then Tim Albert published their answers about Belly Button Fluff (BBF) in an article entitled "Blue jokes — Readers probe the mysteries of the navel".

Peter d'Abrumenil made a telling comment: statistical studies of BBF might not be valid if there is no nudist control group.

Peter Johnson and Geoffrey B. Scott from the Department of Pathology at the University of Aberdeen very cleverly observed that "*abdominal body hair tends towards the umbilicus, as roads to Rome. It is our contention that particles of fluff caught in this bristly trap are cast navelwards under the influence of body movement.*" Of course, this Hair Transport Theory would easily explain why there seems to be less BBF in women that in men — because women have less body hair.

Several readers observed that BBF is not always pure blue in colour, and appears to be related to the colour of the clothing worn.

It's worth noting that blue is a very popular clothing colour in our current society. On most days, most people wear various shades of blue. This might explain why blue is such a common colour in BBF.

PROFESSOR WIL'S THEORY OF BBF

The comedian Wil Anderson (also known as Professor Wil) has his own theory of BBF.

He claims that the body hates colours, and will expel them through the nearest orifice.

So green snot leaves via the nostrils, brown faeces via the anus, yellow urine via the urethra and, yes, blue Belly Button Fluff via the belly button.

EMBRYO vs FOETUS vs BABY

The word "embryo" comes from the Greek "bruô" (or "bryô") meaning "be full of life". This is the name given to the growing baby-to-be for its first two months of life.

From three to nine months the baby-to-be is called a foetus, from the Latin word for "offspring".

ANOTHER THEORY OF BBF

Michael Biesecker also discussed BBF in the 19 April 1995 issue of *Technician* (the student newspaper of North Carolina State University). His theory was similar to Tim Albert's — that the process involves fibres leaving clothes and being funnelled to the belly, where they coalesce into balls of lint.

He wrote that the colour of the BBF is related to the colour of the clothing you wear. But those who wear many different colours usually still generate BBF with the same blue-grey colour that you find in the lint collector of clothes dryers. He says that hairy stomachs are associated with greater quantities of BBF — possibly because the hair both dislodges the fluff from clothing and channels it to the belly button. He also says that larger bellies are associated with more BBF — perhaps because larger bellies have deeper navels.

DOUG SHAVES AROUND HIS NAVEL

I try to avoid opinions, and to stick to the facts. So the lack of hard data would have been the end of the Belly Button question. But then the Soft Bottom Inshore Fish Habitats Research Team (GE, ably assisted by DM) sent me an amazing email about their previously unknown (to me) friend called Doug.

Doug was a Prince Among Men. Instead of just sitting down and thinking, Doug *Had Done the Experiment*.

Doug suffered from/enjoyed BBF. He thought that the hair on his belly was "*channelling and concentrating lint from his clothes into his belly button*". So he decided to test this theory. He shaved his belly over a 10-cm radius around his navel. Suddenly, Doug stopped generating BBF. Doug also noticed that as his belly hair regrew, the BBF reappeared in his navel.

He then made another very interesting observation.

He noticed that the BBF was usually the colour of the underwear below his waist, and not that of his upper body clothing. Perhaps, he suddenly thought, the BBF was being channelled by a Hair Highway (Snail Trail) running upward from his pubic hair to his belly button. This was a reasonable guess. So Doug *Did the Second Experiment*.

BELLY BUTTON vs NAVEL vs UMBILICUS vs OMPHALOS

The belly button has many names, including the fairly technical term "navel". "Navel" comes from the Anglo-Saxon word "nafela".

The Romans called the belly button the "umbilicus".

The Greeks called it the "omphalos". So if you add the Greek word "tomê" (meaning "cutting"), you get "omphalotomy". This word means "cutting of the umbilical cord".

Omphalos also means "knob" or "hub". The Greeks erected a holy stone, or fetish stone, in the Temple of Apollo at Delphi (on the slopes of Mount Parnassus near the Gulf of Corinth). They called this rounded conical stone the Omphalos (or Navel), as they thought that it marked the exact centre of their universe.

The tallest mountain in Bali is Gunung Agung. One Balinese myth says that their deities had mountains as their thrones, and that the highest mountain of all was Gunung Agung. The Balinese call this mountain the "Navel of the World".

The original inhabitants of Easter Island called it "Rapa Nui" ("Great Rapa") or "Te Pito te Henua" ("Navel of the World").

Thar's fluff in that thar belly!

Many good folk have pondered the enigma of Belly Button Fluff (BBF). Some have taken this curious occurrence to even greater heights of thought.

The belly button is our permanent reminder of how we were once connected to Mum by the umbilical cord.

"NAVEL-GAZING"

The phrase "contemplating one's own navel" has the ring of a long and honourable history behind it. The word "omphaloskepsis" (also called "omphaloscopy"), meaning "contemplating one's navel as an aid to meditation", sounds like it is thousands of years old. "Skepsis" is a Greek word meaning "the act of looking, or inquiry". However, the Merriam-Webster web site "Word of the Day" column claims that omphaloskepsis was invented only in the 1920s.

This was not the first time people tried to find enlightenment in the navel. In the past, an "omphalopsychic" was one of a group of mystics who gazed at their own navel so as to induce a hypnotic reverie. The Greek Christian monks of Mount Athos used a specific method of navel contemplation called Hesychasm, to maximise the divine enlightenment. This method would presumably have given them many different insights into divine glory.

But another navel divination method, "omphalomancy", gave only one specific item of information. It predicted how many children a woman would give birth to, by counting the number of knots (bumps in the fleshy plaiting) in her umbilical cord when she was born.

BELLY BUTTON SCAR

Your belly button is your very first scar. It's scar tissue left over from where the umbilical cord joined you to your mother's placenta when you were in her womb. Just like fingerprints, no two belly buttons are alike.

All the nourishment going to the baby and all the wastes coming out passed through the belly button, via the umbilical cord. Once you had been delivered, your umbilical cord was usually clamped or tied, and then cut. The stump withered and fell off after a few days, leaving behind the scar we call the belly button.

Your abdominal wall is made up of various layers, including skin, muscle and fat. They are all fused together at your belly button. You have subcutaneous (literally, "under the skin") fat that plumps up the skin all over your body. But the fat cannot lift the skin at the belly button, because the skin at that location is fused to your abdominal wall. That's why the belly button is concave.

DOUG SHAVES HIS LOWER BELLY

He tested this guess by shaving the hair from part of his Snail Trail, in a horizontal band across his lower abdomen (not from around his navel, as before). This effectively created a hair-free roadblock. Again, the BBF suddenly stopped.

OUR RESPONSE

This unexpected email shamed us into action. If Doug, all by himself, could *Do an Experiment* — then so could we.

Thanks to the Net, it was all so much easier than it would have been 10 years earlier. We set up a BBF survey on my web page, and left it running for two months (see page 96).

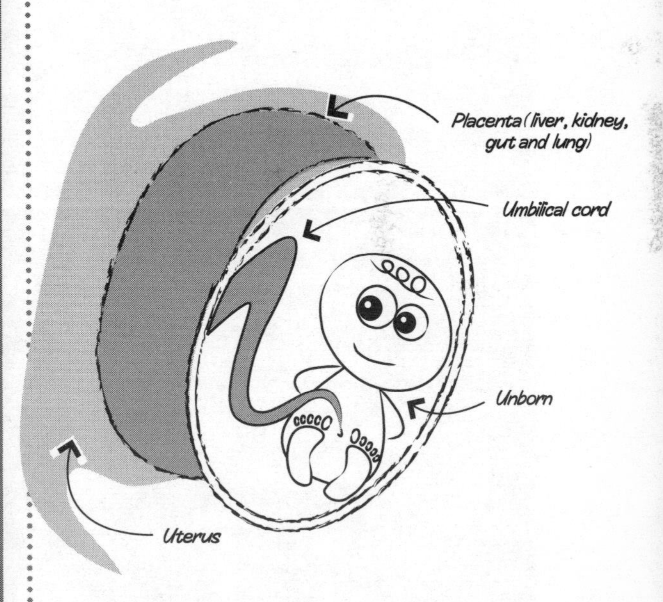

The umbilical cord is made up of four different structures: there are two arteries (taking waste to the placenta), one vein (supplying oxygenated blood and food), the allantois (which degenerates and turns nto the bladder) and the vitello-intestinal duct (which turns into the gut).

PLACENTA AND UMBILICAL CORD

Until the baby is born, it relies almost entirely upon the placenta. The placenta is a strange, flattish organ that acts as a combination of liver, kidney, lungs and intestines.

The placenta supplies the baby-to-be with oxygen and all the nutrients needed for growth. At the same time, it is a barrier that separates the baby from the mother. It grows on the inside of the uterus and lies entirely outside the baby's body. (When I was a medical student, I decided that the uterus was my favourite organ, because of its wonderful design features.)

The placenta is where the baby's blood dumps all its wastes and picks up nutrients. It looks a little like a small, flat cake ("placenta" is the Latin for "cake"). It's about 20 cm across, 3 cm thick in the centre and much thinner at the edges.

The placenta keeps the blood vessels of the mother quite separate from the blood vessels of the baby. However, the blood vessels of the mother and baby run so close to each other that chemicals drift straight through the blood vessel walls. Nutrients travel from the mother's blood vessels to the baby's, and waste products go in the other direction.

Even though the placenta is only the size of a small cake, it has a total surface area of about 13 square metres to ensure efficient exchange of nutrients and waste products. At birth, it usually weighs roughly one-sixth of the weight of the baby — about 500 grams.

But the placenta is also a very hard-working organ. At full term, it makes about 7.5 grams of protein each day. No other organ in the human body makes that much protein.

The umbilical cord is the lifeline that runs from the baby's belly to the placenta. It's a twisted structure about 2 cm in diameter. It increases in length during the pregnancy.

On average, at birth, the umbilical cord is roughly as long as the baby, 50–60 cm, but it can vary between 12 and 152 cm. If it is too long, there is a risk that a loop of the umbilical cord could get caught around the baby's neck as it enters the outside world, strangling the baby in the process. One umbilical cord was so long that it *"looped once around the baby's body, then over the shoulder, under an armpit and twice around the neck, with a good length left over to its root in the placenta"*. If the cord is too short, there can be difficulty in delivering the baby.

Before you were born, the umbilical cord, with its two arteries and single vein, was your lifeline. The power supply to push the blood to and from the placenta was your tiny foetal heart.

The two umbilical arteries carried low-pressure, de-oxygenated blood, loaded with waste products, through your belly button from your body to your mother's placenta. The single umbilical vein carried high-pressure, oxygenated blood, full of nutrients, back in through the belly button. (This is one of the rare cases where arteries carry de-oxygenated blood. Another one occurs in the lungs. Here the pulmonary arteries carry de-oxygenated blood from the right ventricle of the heart to the lungs — and pulmonary veins carry the red oxygenated blood back into the heart's left atrium.)

The blood moves through the umbilical cord at around 6.5 kph. At full term, about 1 litre of blood flows through the umbilical cord every minute.

About one-fifth of babies come out with the umbilical cord around their neck. Modern foetal monitoring techniques can warn the obstetrician or midwife of this occurrence.

After the baby is delivered, it still has the umbilical cord connected to the placenta. The placenta cannot be left inside the uterus — like the baby, it also has to be delivered. If the umbilical cord is too short, as the baby is delivered, it might pull the whole placenta off the uterus before it is "ready" to let go, or tear it apart. The baby could then die from massive blood loss, as its blood drains out through the umbilical cord.

WHAT WAS GOING ON?

Our Working Hypothesis (science talk for "Reasonable Guess") had two parts. We definitely did not think that the belly button generated its own BBF. (However, Professor Wil disagrees — see box, *Professor Wil's Theory of BBF*.)

The first part of our guess was that at least some of the BBF was made from clothing fibres (although there might be other stuff in there as well, such as your own hair, or skin cells). These fibres might fall out because the clothes were getting old and threadbare, or perhaps because your washing machine or clothes dryer was too brutal on your clothes (which is why dryers have lint filters).

The second part of our hypothesis was that the hair on your belly acted like a one-way ratchet mechanism, and advanced the loosened clothing fibres towards the belly button.

UMBILICAL CORD SURVIVES INTERNALLY AFTER BIRTH

The umbilical cord forms in the foetus's second month of life. It contains four main structures, all running through the belly button. They are the pair of umbilical arteries, the single umbilical vein, the allantois, and the vitello-intestinal duct.

Early on, most of the allantois disappears. Inside the baby's body it turns into the bladder.

The vitello-intestinal duct is a structure that ultimately turns into the gut. By the time the growing embryo is just six weeks old, the vitello-intestinal duct has disappeared from the umbilical cord — 98% of the time.

After the baby is delivered, the four structures of the umbilical cord shrink and close up entirely. They turn into internal tendons or cords.

The two obliterated umbilical arteries run downwards to become the lateral umbilical ligaments, which join with the arteries supplying the bladder.

The obliterated umbilical vein becomes the ligamentum teres, which runs upwards and attaches the liver to the belly button.

The obliterated allantois is now a cord that runs down to the tip of the bladder.

OUR EXPERIMENT

A hypothesis is just "spinning your wheels", unless you follow it up with an experiment.

When I was a physicist, we did experiments. When I was a biomedical engineer (designing and building a machine to pick up electrical signals from the human eyeball), this suddenly changed. I was told that research work related to humans was not called an "experiment", it was a "study". My boss told me that was because people get quite anxious when you want to do an "experiment" on them — but not nearly so anxious if you just want to do a "study".

But most people understand what an "experiment" is, so let's call it "Our Experiment". Our Experiment had two parts: the survey, and the practical work.

Our survey asked 19 questions. We even asked the participants to get mega-involved, and shave their bellies to do the Hair-Free Highway Experiment.

Once you've been an air-breathing baby for a while, there should be no connection between your gut and your belly button. But the vitello-intestinal duct can occasionally remain open along its whole length, or just part of its length, up until birth. Very rarely, it remains open past the birth and into adult life.

There are three main outcomes, depending on which section remains open. Sometimes it can discharge mucus or faeces. Sometimes it can form a little cyst behind the belly button. And sometimes it can even form a band that knots around part of the gut, and causes a life-threatening obstruction of the intestine.

About 2% of the population have a Meckel's Diverticulum. It's a little tube located on the front border of your ileum (part of your small intestine). The Meckel's Diverticulum reaches towards the belly button, but doesn't quite make contact.

Bailey and Love's *Short Practice of Surgery* says that the *"umbilicus is a creek into which many ... streams may open ... an enlarged inflamed gall bladder ... may discharge gallstones through the umbilicus. Again, an unremitting flow of pus from ... the umbilicus of a middle-aged woman led to the discovery of a length of gauze overlooked during hysterectomy five years previously"*.

Media release September 4, 2000

Dr Karl Kruszelnicki Needs Your Belly Button Now

Participate in the Great Belly Button Lint Survey

abc.net.au/science/k2/lint
The Lab - ABC Science Online

We know how The Universe began, but we do not yet understand Belly Button Lint.
If you'd like to help us get to the bottom of one of the Greatest Unsolved Mysteries of the 21st Century, then here's your chance. This topic has been a constant and popular discussion thread on Dr Karl Kruszelnicki's Self-Service Science Online Forum and the good doctor has been forced to take matters (and matter) into his own hands, digging deeply to reveal the innermost secrets of belly button lint.

Dr Karl — an Australian Broadcasting Corporation Radio and www star — has been offered various theories by his dedicated, numerous and intelligent audience. However, these theories have also been refuted. In response to the controversy, and powered by our Relentless Search for the Truth, we announce this important survey.

Unzip and log on at abc.net.au/science/k2/lint.

Why is Belly Button Lint always a bluish colour? What causes Belly Button Lint?
Many people believe that the colour of Belly Button Lint is directly related to the colour of clothing — in particular, the colour of undergarments. Check the colour of the lint in your clothes dryer — is it usually a shade of blue? Is the filter in your dryer, in fact, the metaphorical belly button of your home?

Controlled experiments involving shaving, piercing, unusual underwear and vast varieties of fabric have shown some tendencies and trends but a large scale survey is necessary to make meaningful conclusions. These conclusions are important to us all *especially* in this Olympic year in Australia, where some of the finest and fittest navels in the world will be bared as their owners are put through their record-breaking paces.

Dr Karl is available for interviews and you can find out EVERYTHING about him and his extraordinary Self-Service Science Forum Online at abc.net.au/science/k2.

The Lab is the Australian Broadcasting Corporation's Science web site and proudly hosts the online manifestation of Dr Karl Kruszelnicki — including his Thursday morning interactive Science Talkback Online 'n' Radio program on Triple J.

Inquiries: Frankie Lee — ABC Science 61 2 9333 1500

But we also wanted to understand more about the physical structure of BBF. So we asked people to send in samples of their own personal BBF, and then looked at them under various microscopes.

THE TAKE-HOME ANSWER

4799 people answered the survey. The size of this sample was big enough to do reasonable statistics. They were 58.1% male (2790) and 41.9% female (2009). Only 66% (3169) of these people had BBF.

The results can be easily summarised as follows: you're more likely to have BBF if you're male, older, hairy, and have an innie. And the electron microscope image (see back cover) showed typical clothing fibres, with occasional clusters of skin cells. And some tangles that the light microscope showed to be made up of red, white and blue clothing fibres looked blue to the naked eye!

But if you want to find out what happened to BBF when people changed gender, or used navel rings and electric toothbrushes, keep reading. Note that all the following results deal only with the 3169 people who admitted to having BBF.

On the right is a list of the questions we asked. Most of our respondents heard about the survey via Triple J (a "yoof" radio station), but we did get worldwide publicity via the *New Scientist* magazine. Our survey also attracted the attention of the worldwide media, which added to the number of respondents.

SURVEY QUESTIONS

1. First name (optional):
2. Age:
3. Sex:
4. Degree of overall hairiness:
5. Do you have a Snail Trail?
6. Degree of Snail Trail hairiness:
7. Innie/Outie?
8. Describe your build:
9. Skin colour:
10. Skin type:
11. Do you have a navel ring?
12. Do you get Belly Button Lint?
13. If so, what colour is it?
14. Have you noticed that your Belly Button Lint colour is related to the colour of your clothing?
15. If so, what clothing?
16. Do you wash your clothes in a top-loader or a front-loader?
17. Did you undertake the Hair-Free Highway experiment?
18. If yes, did you notice a decrease in Belly Button Lint?
19. Any other comments on Belly Button Lint:

BELLY BUTTON SHAPES

In medicine and surgery, a "symptom" is something that the patient complains of, eg, *"I urinate a lot and I'm always thirsty."* A "sign" is something that the doctor would notice, such as yellowish tissue near the eye.

Hamilton Bailey wrote a famous textbook devoted to signs, *Demonstrations of Physical Signs in Clinical Surgery*. He included many of the signs of the human body that he could describe and/or photograph. He became strangely poetic when he wrote that *"every time an abdomen is examined, the eyes of the clinician, almost instinctively, rest momentarily upon the umbilicus. How innumerable are the variations of this structure!"*

Gerhard Reibmann, a Berlin psychologist, sees the belly button differently from Hamilton Bailey. He believes that you can diagnose a person's life expectancy, general health and psychological state purely by looking at their belly button. He paid for the publication of his own book, which he called *Centred: Understanding Yourself Through Your Navel.*

In it, he reckons that there are six different types of navel. He claims that each one has a specific personality type and a specific life expectancy associated with it. It's easy to be sceptical of something this "easy", although it may turn out to be as inaccurate as phrenology (diagnosing character type by feeling the lumps and bumps on a person's skull).

Gerhard Reibmann, a Berlin psychologist, claims that if you have a horizontal navel (spreading sideways across your tummy), you're likely to be highly emotional, live for only 68 years. But if you have a vertical navel that runs up and down your belly, you'll magically be generous, self-confident and emotionally stable. Somehow, this means that your life expectancy will be around 75 years.

A person with an outie, or protruding belly button, is claimed to be optimistic and enthusiastic and will live for 72 years. However, a person who has a concave, bowl-shaped navel will be gentle, loving, cautious, delicate, sensitive and rather prone to worrying. Presumably, this worry will take a toll on their health, so they'll have the shortest life expectancy of all — only 65 years.

A person with a navel that's off-centre is supposed to be fun-loving and to have wide emotional swings. They're expected to live for only 70 years.

The final (and luckiest) type of navel is the evenly shaped and circular navel. This person is modest and even-tempered and has a quiet, retiring personality — and as a result will live for 81 years.

Now, as we all know, anything to do with the human body always turns out to be more complicated than you first thought. How long will you live if your navel fits more than one of the six categories? Easy, according to Gerhard Reibmann — just add the number of years together and divide by the total number of categories to work out your personal life expectancy.

The average life expectancy in Australia is 83.2 years for women and 77.2 for men. I guess that a lot of Australian women must have navels that are rounder than round.

LITERATURE AND BELLY BUTTON FLUFF

The blue colour of Belly Button Fluff is specifically mentioned in *The Troublesome Offspring of Cardinal Guzman*, by Louis de Bernières. A town is being held under siege by bloodthirsty and cruel religious crusaders. Elders from the town go and ask a mad Englishman, Don Emmanuel, for his advice on how to annoy the crusaders as a form of guerrilla warfare. In his reply, Don Emmanuel speaks of BBF as "dingleberries". Strangely, he admits that he does not perform his own BBF removal, but has Felicidad do it for him ...

"Don Emmanuel grinned, scratched his rufous beard and then his pubic region, and said, 'I will give you all the advice in the world if only you can tell me why it is that the dingleberries excavated from my navel by Felicidad are always composed of blue fluff, when I possess no clothes of that colour.' "

Extract from *The Troublesome Offspring of Cardinal Guzman* by Louis de Bernières, published by Secker & Warburg.

Q2. AGE

We asked people into which five-year age group they fitted, ranging from 16 to 75. The numbers in each age category dropped off evenly with age. There were the greatest number of respondents (768) in the 16–20 group, and the lowest (3) in the 71–75 group. There were still enough in the 51–55 age group (62) to do meaningful statistics.

The clear result was that the older you get, the more BBF you get.

CANCER AND THE BELLY BUTTON

Very rarely, a secondary cancer can be found in the belly button. It's called a Sister Joseph's Nodule, or Sister Mary Joseph Nodule, in honour of Sister Joseph of the Mayo Clinic.

Sister Joseph had an observant clinical eye for patients and their lumps. She had honed it very finely indeed, over a period of very many years. In particular, on a few occasions, Sister Joseph had noticed that a certain type of lump in the belly button would later be associated with a cancer. This cancer would usually be in its late stages.

She told this to Dr William Mayo, who agreed with her. Her "sign" now has a permanent place in surgical history.

Q3. SEX

The majority of those with BBF were male (73%, 2313), while the minority were female (27%, 856).

Males tend to have more hair than females, so this fits our Hair-Causes-BBF Theory.

Q4. DEGREE OF OVERALL HAIRINESS

Hairiness seemed to be related to BBF. About 97% of those who had BBF were either "not very hairy", "moderately hairy" or "very hairy".

Surprisingly for the Hair-Causes-BBF Theory, 3% with "no hair" or "very little hair" also had BBF. Perhaps these people with very small amounts of belly hair wore tight clothes, which helped carry the fluff towards the belly button. Once it fell into the belly button, it wouldn't easily come out again. But perhaps there are some other factors involved . . .

Q5. DO YOU HAVE A SNAIL TRAIL?

About 80% of people who have BBF also have a Snail Trail of hair leading up from their pubic hair to their belly button.

Q6. DEGREE OF SNAIL TRAIL HAIRINESS

In general, women have pubic hair that looks like an inverted pyramid, or a map of Tasmania, with a sharp cut-off at the top. In general, men have pubic hair with a tapering tail of hair reaching up towards the belly button.

However, there is an overlap between the belly hair of men and women — some men have no Snail Trail at all, and

ADAM AND BELLY BUTTON

In the Christian Bible and the Jewish Torah, Adam is the first man and Eve is the first woman.

The existence of Eve is explained in Genesis 2:7 and 21–22, which says: "*7 And the Lord God formed man of the dust of the ground ... 21 And the Lord God caused a deep sleep to fall upon Adam, and he slept: and he took one of his ribs, and closed up the flesh instead thereof; 22 And the rib, which the Lord God had taken from man, made he a woman, and brought her unto the man.*"

Things get even more complicated with the creation of Adam. His belly button gave rise to many philosophical problems.

Some theologians have argued that because he was the first man he had no human parents. Therefore he did not come from a mother, did not have an umbilical cord and did not have a belly button. And surely, they claimed, God would not give us the false impression that Adam (and Eve) came from a mother. But other theologians disagree.

So what was a painter of 500 years ago to do?

Some painters took the easy way out, and covered the belly button area with a strategically placed fig leaf, tree or forearm. But braver painters such as Raphael and Michelangelo gave Adam a navel. In fact when Michelangelo painted Adam on the roof of the Sistine Chapel in the Vatican he gave him a navel — where any worshippers, including the Pope, could easily see it. One of today's radio preachers has condemned Michelangelo as "*immoral and unworthy of painting outhouses and certainly not worthy of painting ceilings*".

Half a millennium later, in 1944, Adam's navel was a problem to a subcommittee of the US House Military Committee (chaired by Congressman Durham of North Carolina). His subcommittee refused authorisation of a 30-page booklet, *Races of Man*, that was to be handed out to American soldiers fighting in World War II. The original booklet had an illustration that showed Adam and Eve each with a navel. The subcommittee ruled that showing Adam's and Eve's navels would be "*misleading to gullible American soldiers*".

It makes you wonder how the soldiers dealt with the horrors of war ...

some women have a small Snail Trail. Most of the people who did have a Snail Trail had only a moderate one.

Our results are a little confusing. It seems that if you have too much belly hair, or too little belly hair, this somehow inhibits the movement of BBF into the belly button. Perhaps a thickly forested abdomen traps the BBF, while an almost bald abdomen doesn't provide enough forward traction.

Q7. INNIE/OUTIE?

About 96% of those with BBF have an innie belly button. Unfortunately, we don't know the ratio of innies and outies in the general population. So we really don't know how to interpret this 96% figure.

Q8. DESCRIBE YOUR BUILD

There was no real correlation between BBF and a person's overall build.

INNIE vs OUTIE

Your normal belly button is concave, with an attractive upper hood. The base of the belly button usually joins onto the muscle wall of the abdomen. Around the belly button there is subcutaneous fat. In the "outie", there is a protuberant mass of subcutaneous scar tissue between the bottom of the belly button and the muscle wall of the abdomen. This scar turns the concave "innie" into a convex "outie".

BELLY TO BREAST

Plastic surgeons are now able to insert breast implants via the belly button. The advantage of this is that it leaves no obvious scar.

The surgeons cut in through the belly button, and insert an endoscope tube under the skin. They work their way over the ribcage until they get to each breast, and then make an opening between the breast and the ribcage. They then insert a rolled-up breast implant into each breast. Once it's in place they fill it with salt water.

Q9. SKIN COLOUR

There was a slight correlation between skin colour and BBF colour. People with darker skin tones had darker fluff. People with lighter skin tones had lighter-coloured fluff. However, the correlation is so small that it is probably within experimental error.

Q10. SKIN TYPE

There was no real correlation between skin type and BBF. About 13.5% said they had dry skin, 70% had "normal" skin, and 15.9% had oily skin.

Q11. DO YOU HAVE A NAVEL RING?

Of the people who have BBF, 92% did not have a navel ring, and 8% did.

There were many fascinating insights in the any-other-comments section (question 19) from people who had a navel ring. Many of them wrote that they did have BBF prior to getting the navel ring. But once the navel ring was installed, the BBF would suddenly either be dramatically reduced, or totally disappear.

It seems to us that BBF was originally transferred into the navel by the Snail Trail, or by friction between the clothing and the belly. The presence of the navel ring might act like a centre pole on a circus tent, and keep the clothing from touching the skin around the navel. So the BBF wouldn't get carried over that last little section to the belly button.

UMBILICAL CORD AND BARBER'S POLE

The umbilical cord has bright red veins spiralling through its white Wharton's Jelly. We see it every day symbolised in the barber's pole.

DRYER LINT IS A CANARY

In the old days, coal miners would take a canary down the mine. Canaries were exquisitely sensitive to some of the dangerous gases. If the canary keeled over, they'd leave the mine.

The lint from your laundry dryer could be a modern-day canary, according to Peter G. Mahaffy from the King's University College in Edmonton, Alberta, and his colleagues.

Back in 1994, the Edmonton Board of Health became concerned about high lead levels in the child of a radiator mechanic. Many of today's car radiators are made of various synthetic plastics. But back then they were made of copper pipes, and fins were soldered onto the pipes using a lead solder. (As the air went over these fins, it took the heat away.) So a radiator mechanic's regular work involved contact with a lot of lead.

Dr Mahaffy realised that lead particles could make their way onto the radiator mechanic's overalls, and then via the family washing machine into the rest of the family's clothes — and into their bodies.

The group tested the clothes dryer lint from radiator-shop employees, and compared it with the lint of people who had no known exposure to lead. The radiator-shop workers had dryer lint with lead levels up to 80 times higher than non-radiator-shop workers.

This is a rather neat screening test for lead. In general, the lead test involves drawing blood, which many children don't enjoy. Screening for lead by examining dryer lint is far cheaper and less invasive.

Qs 12–15. WHAT COLOUR IS YOUR BELLY BUTTON FLUFF, AND IS IT RELATED TO YOUR CLOTHING COLOUR?

About 37% of people with BBF said that the colour of their BBF was related to the colour of their clothing. About half of these people had blue BBF. Most people wear various shades of blue.

But we really can't explain why some people consistently have BBF in a colour that is not present in their clothing.

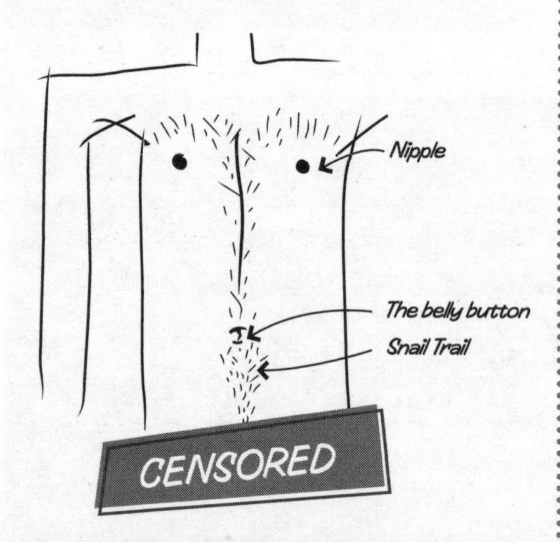

For BBF to form, the hypothesis is that it needs two things: one, clothing fibres (as well as other great ingredients like your own hair and skin cells); and two, hair on your belly. Belly hair acts like a one-way ratchet system directing fibres to the belly button.

Q16. DO YOU WASH YOUR CLOTHES IN A TOP-LOADER OR A FRONT-LOADER?

There is a fundamental difference in how top-loaders and front-loaders treat your clothes. A *Choice* magazine survey showed that in general, a top-loading machine can either be gentle, or can remove a lot of dirt from the clothes — but it can't do both at the same time. Only a front-loading washing machine can be both gentle and thorough.

About 80% of people used a top-loading machine, 16% used a front-loader, and about 3% said they didn't know. Again, we don't know what the percentages are of the different types of washing machines in the general community — so we can't really interpret these results.

However, I have heard many anecdotes that when people have changed from a top-loader to a front-loader, the production of BBF drops dramatically. A typical story is that of a Perth taxi driver. He told me that while he was travelling around Australia for 18 months, and using fairly rough top-loading machines in caravan parks, he had continuous BBF. As soon as he settled in Perth and bought his own

HAIR TRACK DIRECTS JOEYS

Marsupials give birth to their young outside their pouch. The joey (the baby kangaroo) has to find its way to the pouch, by following a "track" in the fur of the mother kangaroo. Inside the pouch lies the source of life, the nipple.

Is nature giving us a clue about the origin of BBF? The joey follows the hairs; does BBF take the same path?

front-loading machine, his production of BBF dropped to zero. Presumably, a non-gentle, top-loading washing machine will loosen up the fibres in your clothing and make them liable to come loose and travel to your belly button.

Qs 17 & 18. DID YOU UNDERTAKE THE HAIR-FREE HIGHWAY EXPERIMENT? IF YES, DID YOU NOTICE A DECREASE IN BELLY BUTTON LINT?

182 people took part in the Hair-Free Highway Experiment. 40% noticed a decrease in BBF levels. 24% did not notice any difference, and 35% didn't know. The 35% unknowns is such a huge percentage that we are not quite sure how to analyse this component of the experiment.

Anecdotally, it seems as though the Snail Trail does have something to do with BBF levels. We need further research.

Q19. ANY OTHER COMMENTS ON BELLY BUTTON LINT

About two-thirds of the respondents actually added some comments.

BBF Observations — Clothes

BBF and colour of clothes:

"I used to work for the New South Wales Fire Brigade (lots of blue clothes). I found even six months after leaving to wear white business shirts five days a week that each morning and evening I still had a navel full of Fire Brigade Blue lint . . . "

BBF and colour of clothes:

"One night after wearing an outfit of pure red cotton for a school play my belly produced a small gem of red lint."

BBF and age of clothes:

"...the fluffiest T-shirts DON'T give the best fluff! For really first-grade fluff production, don an old, worn-out

BELLY BUTTONS, MIDRIFFS AND TOOTHBRUSHES

I had an email from a woman who told me the story of the belly button that infected a man's mouth ...

"Recently my sister went out with her boyfriend clubbing. She was wearing hipsters, and I told her to clean out her belly button, 'cause it was really dirty. She used her electric toothbrush to do the deed.

The problem is that she then used the toothbrush to clean her teeth. She went to kiss her boyfriend, and within a few hours, his tongue started to develop white furry stuff around it. Now, almost a week later, it's still there, only worse. We are very concerned that whatever was in her belly button somehow got transferred to his tongue and mouth area. Could this be possible, as we are very worried?"

"White furry stuff" sounds like a fungus to me. See your local doctor.

and paper-thin T-shirt from the 1970s, get out and lay about 400 square metres of corrugated-iron roofing on a 33ºC summer's day! The prize at the end of the day will bring tears to your eyes! It was like finding a ruby in belly button each night, except it didn't weigh as much, and it felt like satin!"

BBF Observations — A to Z

BBF and activity:

"BB Lint is worse in summer than winter ... perhaps BB Lint is linked to physical activity?!"

BBF and activity:

"I have noticed the more active I am the more I get. Sometimes I have to 'delint' twice a day."

BBF and animals:

"My German short-haired pointer gets brown Belly Button Lint ... please explain why."

BBF and art:

"I predict there will be an Art of Belly Button Lint. I believe with the correct shampoo and styling products we could soon see Belly Button Lint gracing the catwalks in Paris and Milan."

BBF and art:

"I once was doing an art project, and I needed something blue and fluffy for this little weird creature. It dawned on me — I'll use my Belly Button Lint. There was just enough, 'cause my belly button had been collecting it for ages. My art teacher couldn't figure out what it was when I handed in the assignment. He he he."

BBF and changing belly:

"I had my belly button removed recently (for medical reasons — not for the survey), which necessitated shaving the hair between my pubic area up to my breast line. Until the hair grew back, I found that the lint collected between my 'pecs'. When the hair grew back, I had one instance of BBF and it collected in a mass (about the size of a 50c piece) where my belly button used to be. Since then, I have

BELLY BUTTON CLEANERS

A few different people sent in Belly Button Dusters. I didn't know that these devices existed before our survey. They have one job only — to remove BBF from your belly button.

One person sent in a small article from the *South China Morning Post* (10 September 2000) which tells of a different style of belly button cleaner. This Stick-on Belly Button Cleaner is a Japanese invention. It's an adhesive pad which you apply *"over and into the offending area, and then remove it after 10 minutes (making sure you dispose of the evidence discreetly)"*. They're available from the Lung Shing Dispensary Company in Hong Kong at a cost of HK$48 for six adhesive strips.

I reckon it'd be cheaper just to yank the fluff out manually — or you could use a friend's or relative's electric toothbrush ...

found that I no longer have Belly Button Lint at all (I also don't have a belly button!). From my experience, it travels down from the shoulders, however I am yet to discover where it goes when there is nowhere for it to collect in!"

BBF and changing belly:

"I have not had lint since I had twins. Belly is mass of stretch marks and belly button is out of shape."

BBF and fat:

"I can tell when I am putting on or losing weight, because the amount of Belly Button Lint that accumulates increases markedly (and quite suddenly), with a relatively small increase in weight/fatness. Therefore, when I notice that a lot of lint is accumulating, I get off my arse and start going to the gym again. Not a bad indicator, really."

GREEN BELLY BUTTON LINT

Zev Ben-Avi was in the military for 27 years, and is currently an advocate for the Vietnam Veterans Motorcycle Club of Queensland. He wrote to me telling me that *"in all my time, I never saw 'blue' Belly Button Fluff, only green — jungle green, as in issue-type singlets"*.

However, he is not convinced that the fluff comes from the clothing alone. *"To ease the situation with the troops, I found that intellectual activities in the form of apparently inane questions often occupied hours of funny but pointless debate. Obtain three army issue, brand new, jungle green athletic singlets. Weigh them very carefully on a precise machine that will register small but accurate increments. Record these weights on paper and then log the wash, wear and store cycles as they are rotated daily. Every morning and evening, collect and carefully store the Belly Button Fluff that has accumulated.*

After about 12 months, again weigh the accumulated Belly Button Fluff (which is GREEN, not blue) and again weigh the three singlets. The singlets will not have depreciated in weight and the accumulated Belly Button Fluff will approximate the weight of one singlet. The question then remains as to where the 'green' Belly Button Fluff comes from. The questions to be asked are:

1. *If the Belly Button Fluff is not from the singlet, then where did it come from?*
2. *If the Belly Button Fluff comes from the singlets, then why do the singlets not decrease in weight?*
3. *If the Belly Button Fluff does not come from the singlets, then why is it green???"*

I agree with him that BBF still grips tightly onto a few mysteries.

BBF and job description:

"I didn't know about BBL until I got married! It is my job to get the lint out."

BBF and "navel-grazing":

"I find BBL very sexy on the right body — trim, taught. I have had others scrape mine out in a very sexy way using fingers or even tongue."

BBF and personal hygiene:

"I can't touch my belly button, it's too sensitive. Once in a blue Moon I use diluted peroxide to clean it out."

BBF and pride:

"I am the king of Belly Button Lint. I am a one-man factory."

BBF and the Hair-Free Highway:

"I am a M2F Transsexual, and I noticed that when my stomach was waxed, all the fluff disappeared."

BF envy:

"I feel left out because I don't seem to get any lint."

BBF quantities:

"Profuse amounts congregate, as if there's a party in my belly button."

BBF quantities:

"It replenishes itself every six hours."

BBF Uses

BBF storage:

"My friend collects his boyfriend's and stores it in his teddy bear."

BBF clothing:

"I'm saving mine to knit a jacket."

"BBF, yeah yeah":

"I'm collecting it for my male pattern baldness . . . "

BBF homecraft:

"I reckon we should establish Belly Button Lint collection stations, and make doonas and pillows from it. Maybe we could establish a cottage industry, and have people with spinning wheels recreating cotton and other fabrics from the lint."

BBF firestarting:

"It's useful as tinder when out in the wilderness."

BBF lighting:

"Could Belly Button Lint be combined with ear wax to make a candle? This could go some way towards solving the energy crisis."

BBF Qualities

BBF smell:

It was a very common comment that BBF smells pretty bad.

BBF taste:

"It tastes great!"

BBF taste:

"It's yummy!"

BBF taste:

"Belly Button Lint tastes disgusting."

BBF taste:

"Tastes like chicken."

Other F

"I get lint in my brickies' cleavage — could this be construed as the same thing just in a different spot?"

"You've overlooked an important companion to Belly Button Lint — Bum Crack Lint. Lint can also accumulate at the top of your bum crack, near the small of your back. What is the relationship between these two lint accumulators? Are they always the same colour? Does lint migrate from one location to the other? What are the relative lint densities between bum crack and belly button?

There are clearly a number of important, unexplored issues here."

"I don't get Belly Button Lint, but I get Behind-the-Ear Lint and it is always blue in colour!!"

BBF Theory

BBF Theory and washing powders:

"Perhaps the reason why Belly Button Lint is blue is due to the presence of fluorescent materials, common to most washing powders. These fluorescent materials reflect light in the deep blue end of the spectrum. Perhaps spectral analysis of white fibres would reveal that the fibres have been tinted blue (slightly) by repeated washing."

BBF Theory and clothes dyes:

"... clothes that are black usually are dyed with a base colour of blue or brown — which is made really dark. By the time these clothes are ready to shred, or the fibres have worn off to create lint, it has faded to a blue shade rather than a dark blue-black or brown-black. These 'black' items would be washed together and their fibres would get from one garment to another."

BBF Theory and pH balance:

"I recall that an alkali would turn litmus paper blue, and an acid would turn litmus paper red. Could the alkalis in our skin be absorbed by these minute clothing fibres — hence turning the lint blue?"

BBF Theory and sweat:

"I'm no archaeologist, but I remember reading or seeing on a documentary that the Egyptians used to dye things blue using urea. Now, I'm no medical specialist but I seem to remember that sweat contains urea. Therefore I believe that BBF is dyed blue on its way to the navel by the urea in your sweat."

BBF Theory and wax:

"The colour varies, even within one sample. However there appears to be a waxy substance (vaguely similar to ear wax). That makes me think Belly Button Lint is partially made up of dead skin cells sloughed off and caught in the hair around the belly button."

BBF Theory and compost:

"The belly button is a warm environment. Adding some organic material, such as underpants material, you have the perfect place for a body compost bin. I believe the blue tinge of Belly Button Lint is from a natural organic breakdown. The final stages of any compost are not achieved unless the humble worm can devour the raw material to make the nutrients more available. Any compost area void of worms demonstrates a green-blue slimy substance, and

ANIMALS AND BELLY BUTTONS

All mammals have belly buttons. However, in some dogs and cats, they're a little hard to see because they've healed well and they are covered with hair.

usually a toxic aroma. Now, doesn't this describe Belly Button Lint? Add a worm, and every man could grow a tree."

BBF Theory and tattoos:

"I have a tattoo next to my belly button, so the hair to the left is much finer than the other, non-tattooed side. Maybe lint collection is a sideways or circular action with your clothes etc, like an eddy leading lint into your belly button. So if you disrupt the eddy dynamics with a tattoo or ring, the system breaks down and no lint can enter your belly button."

BBF Theory and astrophysics:

"I believe lint is drawn from our underwear by the gravitational force of the tiny black hole that each of us has in our navel."

BBF Theory and human biology:

"I think it actually comes from the inside of your belly. There is a (yet-to-be-discovered) lint gland, which resides just behind your belly button. This works something like our sinuses, except instead of producing mucus, it produces lint."

BBF Theory and retirement:

"The belly button is where lint goes to retire."

GRAHAM BARKER'S COLLECTION

Graham Barker probably has the world's largest collection of Belly Button Fluff. He has been consistently collecting and storing his own BBF since 17 January 1984 — and since then has hardly ever missed a day. He now has three bottles of BBF, generated at 3.03 mg per day, or about 1.1 grams per year. In November 2000, the Guinness World Records officially recognised his collection as the largest collection of BBF in the world.

His BBF does not quite fit the simple theory that BBF comes from the broken fibres of whatever clothes you happen to wear that day. His BBF is always a *"particular shade of red which alters to a yellow-blue after many years [of storage]"*. But surprisingly, he says, *"I almost never wear red clothing, so where does the red fluff come from? Regardless of what colours I wear, the fluff stays red."*

He also claims that the quantity of BBF generated has stayed the same over many years, even though *"looseness of my clothing, the fabric, and the portion of the day my belly is unclothed all vary from day to day"*.

He wisely concludes, *"More research needs to be done to solve this perplexing mystery."*

And behold, we have BBF. It is worthwhile noting that BBF varies from person to person. Key altering factors probably include such things as belly size, hairiness of the stomach, navel depth, gender, age, the nature of the material being worn and many others.

BELLY BUTTON REFERRED PLEASURE AND PAIN

Leanne rang in to my Science Talkback show. She wanted to know why up until a year earlier, whenever she touched her belly button she had felt a pleasurable sensation in her clitoris. Unfortunately, after she had a laparoscopy (which went in via her belly button) she stopped feeling pleasure in her clitoris. In fact, she wondered if she would ever get it back again — because it felt pretty dang good.

The email response was huge. Both Katie and Sharon had had experiences similar to Leanne's. Luckily, their laparoscopies left them with some (but reduced) pleasurable sensation. KF said that she also got pleasurable clitoral sensations when she scratched the lower half of her belly button really deeply. However, KS said touching her belly button made her go to the toilet. J said that the sensation was more painful than pleasurable. L said that scratching her belly button gave her a sensation in her right forearm (but only after she had broken her right elbow).

Greg said that touching his belly button made him nauseous — but only after sex. Jason got a sharp pain in the end of his penis when he scratched his belly button. Rick experienced an unpleasant sensation in his penis while being tattooed around, and partly inside, his belly button.

In reading all the emails, it seemed that most women enjoyed the belly button stimulation, while most men did not.

This seems to be a case of referred sensation. Imagine that both the navel and the genitals send sensation signals to one certain part of the brain. If you stimulate the navel, that certain part of the brain gets the same sensations as if you had stimulated the genitals. However, I have not yet been able to find any references to the nerves of sensation from the navel and genitals being linked in this way. We can only hope that we will become further enlightened in this area.

Philosophy of Science

"I'm glad that Triple J are doing their part to help out with the scientific development of this great country."

"We take much for granted in this world ... Belly Button Lint should be spared from this ignorance. Let us picket the Prime Minister's house, demand research funds and begin a study into possibly the true meaning of existence ..."

An artist's interpretation of magnified BBF

THANKS

I would like to thank the wonderful Kylie Andrews, Bernie Hobbs and Ian Allen at The Lab (ABC Science on-line) for setting up the survey on my web page and running it for two months. And Steven Manos for doing a hefty amount of statistical analysis and deciphering the survey results. I also thank the Electron Microscope Unit (EMU) at the University of Sydney for allowing Steve to have access to their equipment so he could take the beautiful close-up pictures of BBF. And last, but definitely not least, the general public. First, the people who participated in the on-line survey — all 4799 of you. And second, the people who went to the trouble of sending samples of their Belly Button Lint to us, so we could analyse it.

References

Peter Beaconsfield, George Birdwood and Rebecca Beaconsfield, "The placenta", *Scientific American*, Vol. 243, No. 2, August 1980, pp 80–89.

David Bodanis, *The Body Book*, Little, Brown & Company, Boston/Toronto, 1984, pp 126–132.

Peter G. Mahaffy et al, "Laundry dryer lint: A novel matrix for nonintrusive environmental lead screening", *Environmental Science & Technology*, Vol. 32, No. 16, 1998, pp 2467–2473.

Samuel R.M. Reynolds, "The umbilical cord", *Scientific American*, Vol. 187, No. 1, July 1952, pp 70–74.

Corrina Wu, "Dryer lint snares more than just fuzz", *Science News*, Vol. 154, 3 October 1998, p 216.

Room Spins WHEN DRUNK

Why does the room spin when you've had a few too many?

Here's a really bad joke. It goes: "*How many drunk people does it take to change a light bulb?*" And of course the answer is: "*Ten. One person to hold the light bulb, and nine to get drunk enough so that the room spins around the person holding the light bulb.*" And apparently Dean Martin once said, "*You're not drunk if you can lie on the floor without holding* on." All jokes aside, this spinning is a real effect. If a person does get drunk enough, they get the sensation of the room spinning. Why does alcohol do this — and just when you're trying to get some sleep?

BRIEF HISTORY OF ALCOHOL

We humans have been swigging beer for at least 6000 years. The evidence is a chemical called "beerstone", which is rich in calcium oxalate. When you store beer, beerstone slowly precipitates down to the bottom of the container. Archaeologists have found beerstone in the internal grooves of 6000-year-old jars. These jars were in the ancient Sumerian trading post of Godin Tepe, in the Zagros Mountains of western Iran. Some 4000 years later, Godin Tepe later was used as a fortress on the Silk Road.

Shaken, not stirred!

A few hours before the unique feeling of Room Spinning.

One we prepared earlier

It takes about four times the legal driving limit of alcohol to make a room spin. Rest assured, the chap above is exceedingly close to experiencing this.

WHERE ALCOHOL GOES

Alcohol comes in through the mouth. About 20% is absorbed from the stomach into the bloodstream, and about 80% from the small intestine. It passes into the water in the bloodstream (blood is about 60% salt water). This blood alcohol is in a state of dynamic equilibrium. More alcohol may be coming in via the mouth. But to counter this, alcohol is leaving the bloodstream — both via the liver, lungs and kidneys and also straight through the blood vessel walls into the water in the body, outside the bloodstream.

Your body can metabolise and break down about one drink (10 grams of alcohol) per hour. The kidney filters out about 5% of your alcohol, and expels it via the urine. The lungs get rid of another 5% — this can be picked up with a breathalyser unit. The liver breaks down the remaining 90%.

Beer was very popular in ancient Egyptian times. It was like a weakly alcoholic porridge, rather than the clear sparkling brown liquid we enjoy today. Some archaeologists say that the Egyptian pyramids were built on beer. This might explain why they don't have any doors ...

SIMPLE ANSWER

The world has turned a few times since the Sumerian traders drained their jars, but we're still not really 100% sure how too much alcohol makes the room spin. One of the more believable theories claims that it's because the alcohol changes the density of some of the bits of your balance system.

Your ears "hear" by turning sound into electricity. Your ears also contain your balance sensors. Your balance sensors are part of your inner ear system, on each side of your head.

In fact, each inner ear has two separate systems dealing with balance.

BALANCE SYSTEM 1 — SEMICIRCULAR CANALS

The first system deals with rotary movements of your head. This very elegant-looking system looks very similar to a modern gyroscope. It has three circles, called "semicircular canals", joined together at right angles to each other. (The term "semicircular" is a hangover from the days when our micro-anatomy wasn't accurate enough to reveal their true structure.) They pick up rotatory acceleration in each of the three planes — up-down, left-right and backwards-forwards. These circles are hollow and are filled with a liquid.

The hollow circles all meet in an area called the vestibule. Just before they come into the vestibule, they have a swelling called the "ampulla". Inside the ampulla is a ridge called the "crista ampullaris" (sounds like the name of a female teen pop star...). In the crista ampullaris is a blob of jelly called the "cupula".

When your head moves, the liquid in the hollow circle lags behind and pushes the cupula over to one side. It swings back to its original position like a very quick swinging door — in less than 0.1 second. As it gets pushed, tiny "hairs" embedded in the base of the cupula get bent. The cupula is very sensitive to small movements of the liquid.

These hairs stick up from hair cells, which continually emit electricity at a certain frequency. When they get bent one way the frequency increases, and when they bend the other way the frequency drops.

EFFECTS OF ALCOHOL ON SOCIETY

Tobacco is the number one killer drug in Australia. But alcohol comes in at number two — 17% of all drug-related deaths.

Alcohol can harm you in many ways. It is implicated in 44% of fire injuries, 34% of falls and drownings, 30% of car accidents, 50% of assaults, 16% of child abuse cases, 12% of suicides and 10% of machine accidents.

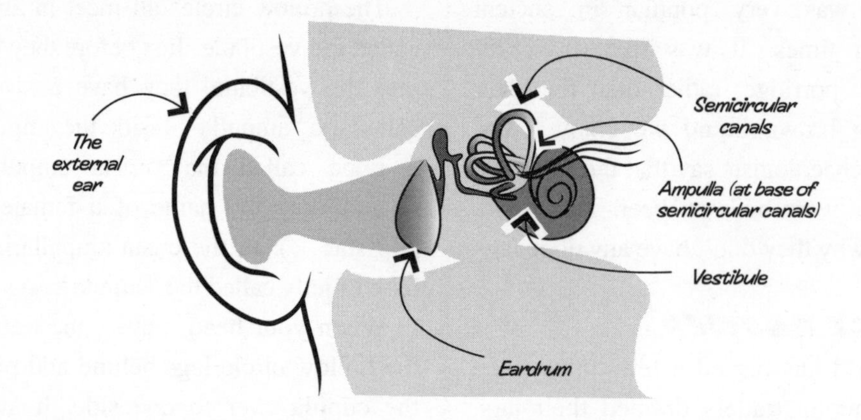

One of the theories that seeks to explain the spinning phenomenon is that alcohol changes the density of some bits of the balance system found within your ear.

THE SILK ROAD

The Silk Road, along which the first alcohol was transported, was not a road. Two thousand years ago, it was a 6400-km-long caravan track that joined the two great civilisations of China and Rome. Silk came to Rome, and in exchange, gold, silver and wool went to China. The Chinese end was started by the Emperor Shih huang-ti around 220 BC.

Very few merchants would travel the entire distance. Instead, they would shuttle back and forth over a short section, trading with other merchants at each end of their territory. It reached its peak use in AD 200. At the time, it was probably the longest road system on Earth.

BALANCE SYSTEM 2 — UTRICLE AND SACCULE

The second system deals with accelerations in a straight line, or the position of your head. Its so-called "gravity receptors" are two swellings called the "utricle" and "saccule". They're located in the vestibule. The utricle and saccule each have a patch of sensory cells called the "macula". Each macula is about 2 mm in diameter. When you stand, the macula of the utricle is in the horizontal plane, while in the saccule it's in the vertical plane.

Just like in the semicircular canals, there are cells with hairs growing out of them. But here the hair cells are not part of an enormous structure like the cupula. Instead, everything is much flatter. The hairs stick up into a thin gelatinous membrane. This membrane is sprinkled with thousands of tiny crystalline stones (called otoliths) made from calcium carbonate. They are about 3–5 microns in size (a scalp hair is about 60–70 microns).

HISTORY OF BALANCE

Back in the 17th and 18th centuries, anatomists thought that the inner ear dealt only with sound. They thought that the three perpendicular semicircular canals constituted a sophisticated device to very accurately locate the source of a sound in three dimensions.

This all changed in 1824, when the French neurologist Marie-Jean-Pierre Flourens did a series of experiments with the semicircular canals in pigeons. He proved they were organs of balance that dealt with the position and movements of the head.

In 1873, Ernst Mach (famous for the Mach number, which measures the speed of sound), Josef Brewer and Crum Brown all independently came up with the same idea. They thought that liquid moving in the hollow semicircular canals triggered the balance signals. Later, the German physiologist J.R. Ewald proved they were correct. In one study on a pigeon, he used a miniature pneumatic hammer to squash one of the canals and force fluid to move. The pigeon turned its head and eyes to the opposite side.

These stones are about three times denser than the gelatinous membrane. They are heavy enough to make sure that the hairs bend when you tilt your head or accelerate. They help generate a shearing force between the gelatinous membrane and the hair cells. This in turn makes the hair cells give off electrical signals. If you are moving at a constant speed (like a passenger in a car) then the utricle and saccule don't trigger. You have to rely on other signals (like looking out the window) to work out what your motion is.

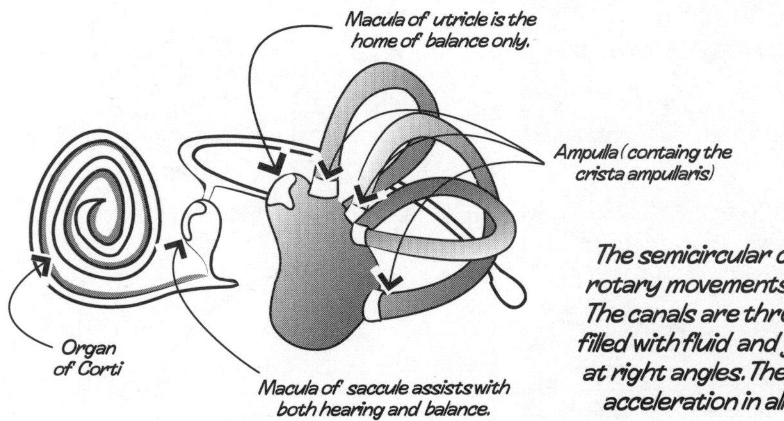

The semicircular canals deal with rotary movements of your head. The canals are three hollow circles filled with fluid and joined together at right angles. They detect rotary acceleration in all three planes.

BALANCE = LITTLE STONES

We have learnt a lot about balance from experiments done with various sea creatures, such as the crab, shrimp and octopus.

The "simplest" balance system of all just has a grain or rock in a hollow chamber — clams have this. It's entirely surrounded by hairs that all point towards the centre. The stone rests on a few hairs and triggers them into firing. As the sea creature changes its orientation, the stone shifts to rest on different hairs, which then send signals to the brain.

In some crustaceans the hollow chamber is still open to the sea. When they moult their old hard shell, they also get rid of the grain of sand. They then pick up a new grain of sand from the environment.

In a famous experiment, scientists put tiny iron particles in the water. A shrimp incorporated an iron particle into its balance organ. When the scientists brought a magnet close to the shrimp, its sense of direction got re-oriented. It now thought that the magnet, not the ocean floor, was "down".

SUMMARY OF BALANCE

The two balance mechanisms (the three semicircular canals and the utricle-saccule pair) have a basic similarity. In both systems tiny hairs bend and generate "balance" signals. These signals travel via the VIIIth Cranial Nerve into your brain — and suddenly you have the sensation that you're moving.

So why does the room spin when you drink too much? Because a "false" balance signal gets sent to the brain.

Inside the crista ampullaris

Alcohol gets into the blob of jelly called the cupula and changes the cupula's density. This distorts its shape, and bends the little hairs embedded in the cupula's base. This results in the hairs giving incorrect electrical signals to the brain to say that you are moving, and leaves you with that unwanted familiar sensation of Room Spinning.

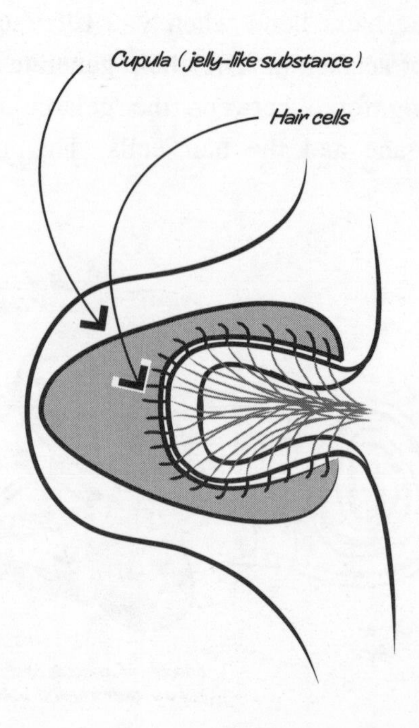

WHAT ALCOHOL DOES

The effects of alcohol depend mostly on the BAC (blood alcohol concentration). In Australia, the legal BAC for driving a car is less than 0.05%. In Sweden, it's under 0.02%.

BAC: 0.03– 0.12%.
Euphoric, self-confident, but trouble with emotional judgment and fine motor movements such as writing.

BAC: 0.09– 0.25%.
Excited but may be sleepy and uncoordinated, may have blurry vision, slower reactions, trouble with understanding and sensing surroundings.

BAC: 0.18– 0.30%.
Confused, sleepy, uncoordinated, dizzy, staggering, very emotional, very blurry vision, doesn't feel pain very well.

BAC: 0.25– 0.4%.
Stupor, barely conscious, doesn't respond, can't walk or talk.

BAC: 0.35– 0.5%.
Coma, slow breathing and heart rates, lower body temperature, depressed neurological reflexes (lessened gag reflex, pupils don't constrict to light, etc).

BAC: >0.5%.
Usually dies.

However, it's not all bad news. Alcohol has many good effects on your health. But you must avoid binges, and keep your BAC under 0.5%. As Benjamin Franklin said, *"Beer is proof that God loves us and wants us to be happy."*

EARLY EXPERIMENTS AND JERKY EYES

Back in 1842, the French neurologist Marie-Jean-Pierre Flourens got some animals drunk and saw their eyes doing PAN (positional alcoholic nystagmus). Nystagmus is an involuntary movement of the eyes. It can by left and right, up and down or circular. Pendular nystagmus is when the movements are rhythmical and even. Flourens observed jerk nystagmus — where the movements are quicker in one direction than the other. If the quicker movement is to the left, you call it left nystagmus; if it is to the right, you call it right nystagmus.

When you spin around, you get jerk nystagmus as your eyes flick left and right. Doctors check for this in newborn babies. They hold the baby so that the face is level with theirs, facing them, and about 30 cm distant. Not too rapidly, they spin around a few times while looking at the baby's eyes. If the baby's balance system is intact and wired up correctly, the doctor sees the baby's eyes flicking left and right in a jerk nystagmus.

So there's a relationship between your balance centres and nystagmus.

Since Flourens, many neurologists have looked at the relationship between PAN and alcohol in humans. This research gives us more clues as to exactly where in your body alcohol affects your balance centres.

JERKY EYES AND ALCOHOL

Jerky eyes can be observed when people reach a level of 0.25–0.4% BAC (blood alcohol concentration), which is way above the legal limit of 0.05%.

After about 30 minutes, they'll start showing a jerk nystagmus. You can see the bulge of the cornea flicking back and forth under their closed eyelids. PAN I lasts for about three hours, as the BAC continues to rise.

SPEAKING MORE LOUDLY WHEN DRUNK

Why do people speak more loudly when they're drunk?

The answer is that the alcohol slightly anaesthetises the VIIIth Cranial Nerve, which deals with both hearing and balance. This reduces the level of the audio signals sent to the brain.

The reason why you speak at a certain volume is because of a feedback loop. You monitor how loudly you sound to yourself, and then adjust the loudness of your speaking voice.

So when you drink alcohol you "hear" yourself speaking softly, and then you crank up your speaking volume.

PARTY ANIMALS

In general, women get higher blood alcohol levels than men do. This holds true even when you compare men and women of the same height, weight and general build who drink the same amount of alcohol.

There are two reasons for this:

1. First, men tend to have more muscle and less fat than women. Muscle contains more water than fat. So men have slightly more water than women of identical weight. The alcohol can be diluted in a greater volume of water, which leads to a lower blood alcohol level in men.
2. The second reason is that women don't have a certain key enzyme in their stomachs. This enzyme (glutathione-dependent formaldehyde dehydrogenase) is one of the many that break down alcohol. Consider a man and a woman of identical weight and height who have the same amount of alcohol in their stomachs. The man has the enzyme. He will break down the alcohol more quickly, and end up with a lower blood alcohol level.

But there's one important proviso to this. This scenario holds only if you are drinking beverages containing 10–40% alcohol (wine, champagne and spirits). With high-alcohol drinks like these, the enzyme is activated in men but not in women. But the enzyme is not activated in either men or women when they drink 5% alcohol (such as beer).

There are a few lessons from this:

1. If women want to keep up with the men, they should stick to beer. (The trouble with this advice is that alcohol affects women's organs, such as the liver, more than it affects men's organs.)
2. Men have this enzyme in the stomach and in the liver. Women have it only in the liver, not the stomach. Alcohol gets to the liver via the bloodstream. So if the women want to keep up with the men while having wine and spirits, the only way they can do this is to take alcohol the dangerous way — intravenously. (Don't do this at home, folks.)

Seriously, the main lesson is that men and women should drink out of different glasses — bigger ones for men, and smaller ones for women.

SHAKESPEARE RIGHT AGAIN

Macduff asks the porter in Act 2 Scene 3 of *Macbeth*:

"What three things does drink especially promote?"

Porter: *"Marry sir, nose-painting, sleep, and urine."*

"Nose-painting" refers to the flushed red nose. *"Sleep"* happens because alcohol depresses the central cortex, then the limbic system, the cerebellum, the hypothalamus and pituitary gland, and finally the medulla (in the brain stem).

Urine production is increased above its usual 1 ml per minute. The pituitary gland makes ADH (anti-diuretic hormone), which tends to reduce the output of urine. Alcohol depresses the pituitary gland, which means less ADH and more urine.

Shakespeare was right again ...

As the BAC reaches its peak, the nystagmus vanishes. This quiet period lasts a few hours.

Finally, as the BAC starts falling, PAN II appears. It's just like PAN I, except that the fast flick is in the opposite direction. PAN II lasts for several hours after the BAC has dropped to zero.

THE FULL ANSWER

It seems that alcohol (which is a terrific solvent) gets into your bloodstream, and then into the cupula. The cupula changes its density. This distorts its shape, and makes the little hairs bend. The hairs then send an electrical signal to your brain indicating you are undergoing some kind of rotary acceleration — or in plain English, that the room around you is spinning. This effect is stronger when you are on your bed in a darkened room, and there are no other signals (such as from your eyes) to tell you that the room is, in fact, perfectly still.

The direction in which the room spins depends upon which way your head hits the pillow.

As the night wears on, the alcohol gradually diffuses out of the little cupula. The electrical signals slow down and the sensation of spinning lessens. After a few more hours, the alcohol has almost entirely left the blob of jelly. But your brain has spent several hours thinking that your head is spinning. It has got used to it, so it interprets NOT spinning as seeming to spin in the opposite direction.

So you could go to sleep with the room spinning one way, and wake up with it spinning the other way — and on those days, you're more likely to get out of bed on the wrong side ...

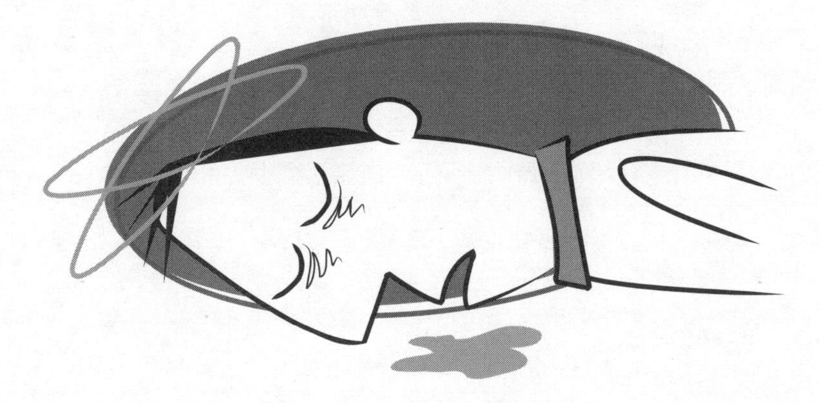

Now comes the enjoyable slumber ...

LOUD MUSIC IS FUN

We're not sure why some loud music is fun. But Neil Todd, a psychologist from the University of Manchester, says that it's related to a part of our hearing system that we have inherited from fish.

In fish, the saccule is used for hearing. In humans, the saccule is located at the base of our hearing device, the Organ of Corti (see diagram, page 117). But most scientists don't think that the saccule has any hearing function, only a balance function.

Neil Todd did a study with 11 university students (a rather small sample size, unfortunately). He found that their saccules would give off electricity in response to sounds between 50 and 1000 Hz. The peak sensitivity was around 300–350 Hz, which is very close to the central frequency of the human voice.

The saccule sends signals to many parts of the brain, including the hypothalamus. The hypothalamus has many jobs — one involves your desires for food, sex and pleasure.

However, the saccule responds only to music louder than 90 dB. Maybe loud music stimulates the saccule into telling the hypothalamus that you're having a good time ...

ANOTHER REASON WHY YOU FALL OVER

Too much alcohol can depress the cerebellum, interfering with its job of coordinating your movements. This can make you stagger and fall over.

But there's another reason why alcohol can make you fall over.

If you stand up suddenly your blood pressure will drop. If you are sober your blood vessels will immediately constrict, pushing your blood pressure back up to normal levels.

But if you are drunk, the blood vessels don't constrict. Your blood pressure stays low — in some cases, low enough to deprive your brain of blood. You faint. But once your body is horizontal, your brain gets enough blood and you recover.

References

Gunnar Aschan and Martin Bergstedt, "Positional alcohol nystagmus (PAN) in man following repeated alcohol doses", *Acta Otolaryngol Supplement*, Vol. 330, 1975, pp 15–29.

G.M. Halmagyi and P.D. Cremer, "Assessment and treatment of dizziness", *Journal of Neurology, Neurosurgery & Psychiatry*, Vol. 68, No. 2, February 2000, pp 129–134.

Paul Marks, "Blast from the past", *New Scientist*, No. 2226, 19 February 2000, p 11.

Alison Motluk, "Why drunken binges end in the gutter", *New Scientist*, No. 2225, 12 February 2000, p 20.

W.J. Oosterveld, "Effect of gravity on positional alcohol nystagmus (PAN)", *Aerospace Medicine*, May 1970, pp 557–560.

Smell AND MEMORY

How can smelling an odour from your distant past (such as chalk dust, or the perfume of your first lover) unleash a flood of memories that are so intense and striking that they seem real?

This kind of memory, where an unexpected re-encounter with a scent brings back a rush of memories, is called a "Proustian memory". It's named after Marcel Proust, one of the greatest novelists of the 20th century. He describes the power of smell in the opening chapter of *Swann's Way*, the first volume in his mammoth seven-part work, *Remembrance of Things Past*.

He writes how the smell of a madeleine cake (a small, rich, cookie like pastry), dipped into lime-blossom tea, brought back crystal-clear memories of his youth: "*and as soon as I had recognised the taste of madeleine soaked in her decoction of lime-blossom which my aunt used to give me ... immediately the old grey house ... rose up like a stage set ... and with the house the town ... the Square where I used to be sent before lunch, the streets along which I used to run errands, the country roads we took when it was fine.*"

THE SENSES

Traditionally, we humans have five senses: smell, hearing, vision, touch and taste. Only two of these senses (smell and taste) sample the chemicals around us for information.

The sense of smell is probably the first sense to evolve in a living creature. Back in the early days of evolution, when we began as single-celled creatures, our sense of "smell" told us what was safe to eat. All living creatures have a sense to detect chemicals in their immediate environment.

In the more complicated animals, the sense of smell is used for other aspects of behaviour, such as finding a mate, synchronising menstrual cycles and communicating with the other animals in your group. A breast-feeding baby can distinguish the smell of his or her mother from another nursing mother. Dogs and horses can smell fear in humans.

The average person can recognise about 4000 different scents, but a trained nose can identify 10 000. People who were born deaf and blind can recognise rooms and people by their smells alone.

SMELL vs TASTE

The sense of smell is more sensitive than the sense of taste.

Strychnine affects your sense of taste very powerfully — it can be picked up in concentrations as low as a billion billion molecules per litre. Mercaptan is a similarly powerful stimulant in the land of smell — it can be detected as low as 10 000 million molecules per litre.

Now, it's a little hard to compare smell and taste sensitivities. To identify a smell you have to sniff a large volume of air, but to identify a taste you need only a very small amount of mashed-up solid or liquid.

However, when you take account of relative volumes, smell is 10 000 times more sensitive than taste.

"INTRODUCTION" TO SMELL

Our sense of smell starts high up in our nostrils, where chemicals land on cells. These cells in the roof of the nose send electrical information to the two olfactory bulbs just inside the skull, and then the signal splits. It finally ends up in both the (modern) frontal cortex and the (ancient) limbic system in the brain.

We humans "smell" with a yellowish area in the roof of each nostril, just underneath and between the eyes. It's called the "olfactory epithelium". It's out of the main airflow, so it doesn't usually get good ventilation. So when you sniff deeply to better appreciate a faint odour, you blast extra air over this yellow patch.

The two patches have a total area of about 5 sq cm — not much, compared with dogs at 18 sq cm or cats at 21 sq cm. So ideally we should have blood-cats instead of blood-hounds — but we use dogs probably because cats are very independent and don't like to be trained. (Actually, catfish would be even better, as they are just one giant olfactory receptor — but they would work only in the water, not on dry land.)

There are two types of cells in the olfactory epithelium: the receptor cells that do the actual "smelling", and the Bowman's Glands cells. The Bowman's Glands cells generate a thin layer of mucus (about 60 microns thick — roughly the thickness of a human hair) on the olfactory mucosa.

Your smell is familiar to me!

How can smelling an odour unleash intense memories from your past?

A SHORT HISTORY OF MARCEL PROUST (1871 – 1922)

Proust's mother was pregnant with him during the Franco-Prussian War, and during the period when a socialist commune was briefly set up in Paris. These were very harsh times. Paris was besieged, and coal and wood ran out so that the houses couldn't be heated during the bitter winter. The Parisians were so hungry that they not only ate their cats and dogs, but even the animals in the zoo.

His mother, Jeanne Proust, was so frail from hunger that when Marcel was born he was so weak and sickly that he was not expected to live.

After his father and mother died in 1903 and 1905 respectively, he withdrew from the world. In his early days of writing, many people derided him as a shallow society writer. He lived as a virtual recluse in a special sound-proofed room until 1919. He had it lined with cork, because he hated noise.

"CHEMISTRY" OF SMELL

All of the chemicals that we can smell must dissolve in this mucus layer. This mucus is rich in fats, mucopolysaccharides, immunoglobulins, proteins such as lysozyme, and various enzymes such as peptidases. These chemicals in the mucus layer make sure that the incoming smell chemicals get presented, or shown, to the receptor cells (also called olfactory neurones).

We can't smell all the chemicals in our environment. They have to be soluble in both water and fat. They also need a molecular weight of less than 300, so that they are light enough to

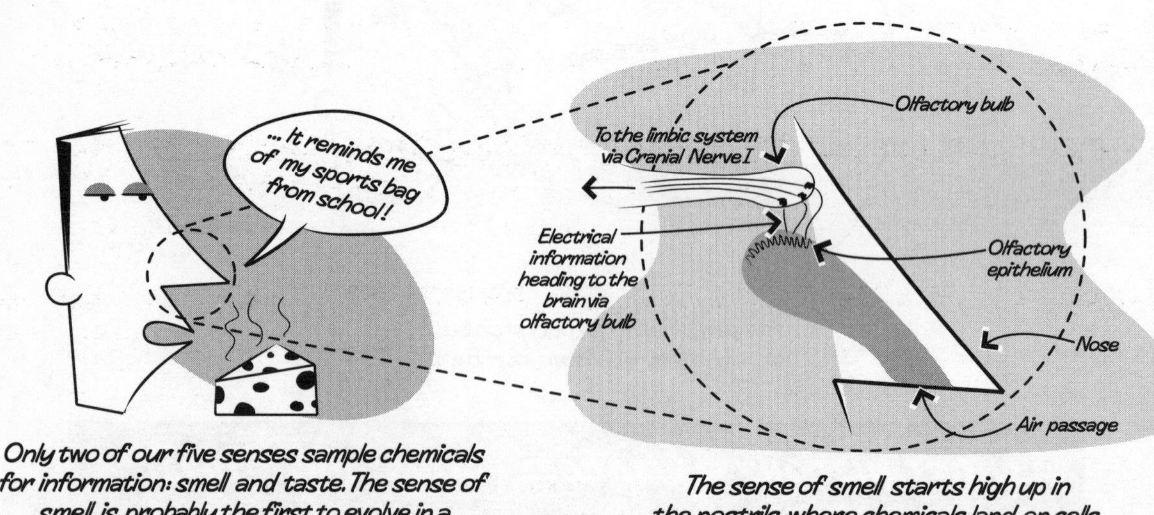

Only two of our five senses sample chemicals for information: smell and taste. The sense of smell is probably the first to evolve in a living creature.

The sense of smell starts high up in the nostrils, where chemicals land on cells.

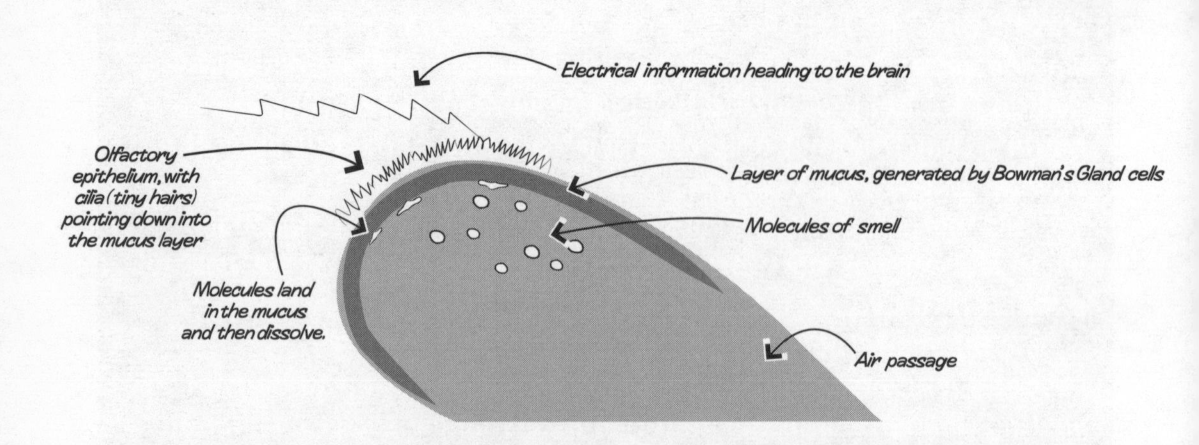

The cells in the roof of the nose send electrical information to the two olfactory bulbs just inside the skull ... then the signal splits. It ends up in both the "modern" frontal cortex and the "primitive" limbic system—the place that is excellent at recognising a smell and then linking it to a past familiar situation.

vaporise, travel through the air and land on the mucus layer in your nose.

A molecule with these properties can dissolve in the mucus layer, and then stimulate the 10 million or so receptor cells that we humans have. Actually, they don't touch the main body of the cells, but instead they land on tiny hairs that sprout off the receptor cells. These hairs are called "cilia". The receptor cells are in the olfactory epithelium, while the cilia reach into the layer of mucus. There are about 8 to 20 cilia per receptor cell. The cilia themselves are between 30 and 200 microns long.

These receptor cells are actually nerve cells. In general, in the rest of the body, once a nerve cell dies it's gone forever. But these olfactory receptor nerve cells are special, and can divide and regenerate themselves every 40 days or so.

"ELECTRICITY" OF SMELL

The receptor cells are where the action happens — where chemicals are "sensed", and in response, electrical signals are generated. These 10 million or so receptor cells send their electrical signals to about 50 000 cells in the two olfactory bulbs, just inside the skull. This "shrinkage" from 10 million to 50 000 has the effect of increasing the sensitivity of our smelling sense.

Once the nerves leave the olfactory bulb, they travel to two main destinations. One destination is the thalamus and the frontal cortex of the brain — so the animal can consciously recognise that it has detected a certain smell. The other

LOUSY HUMAN SMELL

There are genes in the human DNA that deal with processing smells. But most of them seem to be broken, in that they have mutations that have disabled them so they don't work.

At least, that's what Dominique Giorgi from the Institute of Human Genetics in Montpellier, France, and colleagues found. They looked at the genes that make the proteins (in the cilia of the receptor cells) that actually touch, and react with, incoming smell chemicals. They compared these genes in humans and the other primates. They found that the closer a primate was to a human, the more broken genes that it had.

In baboons, 19% of these genes had disabling mutations. The gorillas came in at 50%, while humans topped the list at 70%.

Maybe that's why we are so bad at picking up smells, and so insensitive to the world around us.

PERFUME HUNTERS

Back in Roman times, the perfumers were called "unguentarii". They mixed ingredients brought to them by the perfume hunters.

Today we still have perfume hunters. The results of their work are seen everywhere from perfumes to washing powders, and in that unique "new car smell". They try to find fresh and natural smells, like "waterfall", and perhaps Kramer's "beach" (from an episode of "Seinfeld"). They look in places like Madagascar, an evolutionary hot spot in which 85% of the flowers are found nowhere else on the planet.

Their tools are simple. Once they find a flower with a new smell, they place a glass jar over it — this is sometimes very difficult in a rainforest. The trapped fragrance molecules are then passed over a series of filters, each of which absorbs different chemicals. If it's too difficult to trap the chemicals with a glass jar, they suck air through a hollow tube of silicon rubber.

destination is the limbic area of the brain — an ancient region from the earliest stages of evolution that deals with emotion, pleasure, motivation and types of memory associated with food.

From reading this little description, you would think that we have a good understanding of smell. But we don't. For one thing, in experiments, it's very difficult to guarantee that you can consistently deliver the same number of smell molecules to each nostril of a person. And we still don't know exactly how a chemical landing on a nerve generates electricity in that nerve (see box, *Five Theories of Smell-to-Electricity*).

SMELL IS "PRIMITIVE"

Whichever way it happens, electrical signals get generated in the receptor cells and do go to the limbic area of the brain. Now, here's a really weird thing. The limbic area of the brain evolved directly

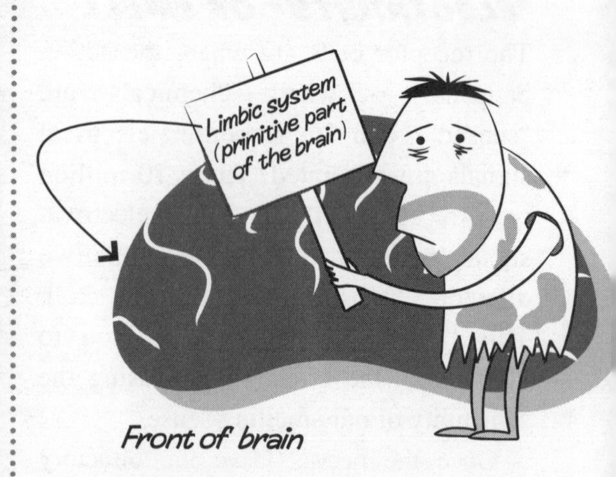

We call that memory which is unleashed by a smell a "Proustian memory". We think this happens in the limbic system, the "primitive" part of the brain.

from primitive smell structures. According to Professor Rachel S. Herz, "*the ability to experience and express emotion grew directly out of the brain's ability to process smells*".

Rachel S. Herz is Assistant Professor of Experimental Psychology at Brown University in Providence, Rhode Island. She's a scientist who specialises in understanding the connections between smells, emotion and memory. She explores how smells can trigger memories, how emotions influence what you smell, how smells in turn link to emotions and behaviour, how the environment in which you experience smells affects your perception of them, and how men and women pick up and process smell information differently in their sexual strategies.

THE ULTIMATE NOSE

Gypsy moths have the best sense of smell ever measured. In fact, it cannot be bettered. They are so sensitive that they can pick up the smell of a single molecule of the moth pheromone called bombykol.

CHEMICALS AND RECEPTOR CELLS

The sense of smell is different from the sense of vision.

In vision, there are three types of receptor cells that handle colour vision in daylight. They are sensitive to blue light, green light and red light. There is some overlap, so that some colours of incoming light will affect (say) both blue-sensitive and green-sensitive receptor cells.

But in smell, we don't know how many types of receptor cells there are. Linda Buck from Harvard University found a family of 1000 or so genes, each of which deals with a single smell receptor.

A single type of receptor can recognise many different types of odours. And many different types of receptors can respond to a single odour. And finally, different odours are recognised by different combinations of receptors. It's a bit like a "language", where a small number of letters are used in different combinations to make a huge number of words.

And even how you sniff can change the final "smell" of the odour. If you have a very gentle sniff, the odour will land on only a few receptors. But if you take a vigorous snort, the odour can stimulate many different types of receptors. This might help explain why a gentle sniff of the chemical called "indole" smells flowery, while a healthy snort makes indole smell like faeces.

FIVE THEORIES OF SMELL-TO-ELECTRICITY

The most popular theory reckons that there's a "lock and key" situation. It's officially called the Molecular Shape Theory. Here the incoming chemical is the "key", which fits into a "lock" on the nerve. Once the "key" fits neatly into the "lock", an unknown reaction then triggers the nerve to generate some electricity.

The second theory is the Diffusion Pore Theory. This theory says that the incoming chemical slowly diffuses or punches its way through the membrane of the receptor cell. As it does so, it leaves behind a hole in the cell membrane. This then somehow generates electricity.

The Piezo-Effect Theory is a little weird. The olfactory epithelium is yellowish in colour, because of a family of chemicals called "carotenoids". This smelling area in humans is pale yellow, but is a much darker yellow in the animals that have a more sensitive sense of smell. The Piezo-Effect Theory says that these carotenoid chemicals combine with the smell chemicals, to generate a weak electrical current.

The fourth theory is the Molecular Vibration Theory. All chemicals have a natural resonant vibrational frequency. Chemicals of smell are all quite small, so their natural frequencies are in the infrared band. This theory claims that the olfactory nerves are directly sensitive to these tiny vibrations, which somehow generate electricity. Of course, different frequencies would give you different smells.

The fifth and final theory is called the Nose-as-a-Spectroscope Theory. This theory claims that when a smell chemical lands on a receptor cell, some kind of strange quantum "electron tunnelling" happens. This tunnelling then forces a cascade of chemicals to be released inside the receptor cell, which eventually leads to electricity being generated in that receptor cell. The "spectroscope" link comes in when it's claimed that different receptors are simultaneously tuned to different smell chemicals.

SMELL AND MEMORY

Professor Herz has been working in the Land of Smell for over a decade. She agrees entirely that the Proustian memory (where a smell boots a realistic mental image into existence) is a real phenomenon. She also agrees that the memories can be very emotionally powerful, and vivid as well. But she has done various experiments that show that the memories triggered by smell are not more accurate than memories triggered by other cues (such as words or pictures). This is a real surprise.

In other words, the sensation that these Proustian memories are very accurate is an illusion, created by that powerful rush of emotion.

Now, how does a Proustian memory work? Professor Herz suggests a few mechanisms. She reckons it's because the smell signals from the nose go to the

PONG WARFARE

The Pentagon wants to use horrible smells as part of its Nonlethal Weapons Program. Bad smells could drive away a hostile crowd of protesting civilians, set up a no-go buffer zone between hostile factions, mark bank robbers so that they couldn't disappear into a crowd, or even end a siege.

Back in World War II, the US Office of Strategic Studies wanted to invent a chemical that the French Resistance could spray onto German officers to humiliate them. It had the charming name of "Who Me?". The chemical was invented, but it could not be delivered "surgically" to the target — everybody in the area, including the deliverer, got tainted with the noxious smell.

One problem today is that a smell can be disgusting to one person, but pleasant to another, eg, horse manure. But one smell that is repulsive to everybody so far tested is "US Government Standard Bathroom Malodor". It was originally invented to test how well bathroom cleaning products were doing their job. Mixing "Bathroom Malodor" with an updated "Who Me?" might work on everybody.

Pam Dalton, a cognitive psychologist working at the Monell Chemical Senses Center in Philadelphia, is trying to find the "correct" ratio of "Bathroom Malodor" and "Who Me?". But success would have a personal price. She worries that if she succeeds, *"...I'm just not sure that I could keep anyone here working with me"*.

limbic system, which deals with memory, basic emotion and motivated behaviour. She also reckons that a lot of memory, smell and emotional processing happens in the right half of the brain.

When you're making or laying down a memory, you mainly store it in the left dorsal prefrontal cortex. But when you try to retrieve that memory, that activity happens in the right prefrontal cortex. Another processing activity that happens in the right brain is trying to work out what an odour or smell is. And when you try to recall an emotional event that happened to you, that also mainly happens in the right brain — in a part of the right limbic system called the amygdala-hippocampal complex.

So these three things (emotion, retrieval of memories, and the sensation of smell) all happen on the right side of the brain. They are also, thanks to our evolutionary history, very closely wired together. Because of this, the sense of smell has a very direct connection with our subconscious minds and memories. This begins to explain how an unexpected smell can bring back clear memories of a past event.

AIR-CONDITIONING NOSE

Apart from Cyrano de Bergerac, the nose does not have much of a romantic history. Thousands of poems praise the eyes, but very few (if any) have ever admired the nose. All that happens to the nose is that you either pay through it, or you keep it to the grindstone.

Every day the nose deals with about 10 000 litres of air. This air has to be cleaned, warmed and humidified to the climate of a hot, humid summer day (about 37°C, with 80% humidity), so that it won't shock the delicate lungs.

The cleaning system is simple, but efficient. Coarse hairs inside the nose strain out most of the particles bigger than 10 microns (about 1/7 the diameter of a human hair). Particles smaller than 10 microns slip past the hairs, but get caught by sticky mucus membranes inside the nose and on the tonsils and adenoids.

The second air-conditioning job of the nose is to warm up the incoming air. This is done by running the incoming air past three hanging bulbs (called "turbinates") of erectile tissue. This is similar to erectile tissue in the penis and clitoris. These erectile tissues can inflate, which gives a greater surface area for the incoming air to kiss against. This transfer of heat is so efficient that when you

USEFUL SMELL RESEARCH

Professor Herz's work has many potential uses.

It might help us understand more about the reliability of a witness in a court case. Unfortunately, witnesses are not always reliable. Witnesses have occasionally been convinced that their memories are truly accurate — but later events have shown that they were genuinely misguided. As Professor Herz has shown, a memory triggered by an odour is no more accurate than a memory triggered by a description, even though the odour memory is far more vivid and seems to be "real".

Another important area of research is to look at how people who are born without a sense of smell experience emotion and memory differently from people with a regular sense of smell. We still don't really understand emotion and memory.

Professor Herz's work might also help people who have lost their sense of smell, either through injury or disease. These people say that they get a progressive blunting of all of their emotions, and often suffer depression.

So this is indeed valuable research, not to be sniffed at . . .

breathe in air at 0°C through your nose, it has warmed up to body temperature by the time it hits the back of the throat.

The turbinates have another job as well. They help you wake up relaxed and fresh in the morning. When you sleep on your left side, the turbinates in your left nostril become engorged and swollen with blood. This sends a signal to the brain, which makes you turn over while you are still sleeping, so that you don't wake up all stiff and cramped in the morning.

To humidify the incoming air, the nose secretes about 1 litre of moisture a day — mainly as a sticky mucus. This mucus acts like sticky paper to trap tiny bacteria and particles. But if these bacteria were allowed to stay in place, they would rapidly multiply and give you an overwhelming infection. To stop this from happening, cells at the back of the nose are equipped with tiny cilia — miniature hairs. The cilia operate at about 10 strokes per second. These cilia push the film of mucus to the back of the throat, where you swallow it. In the stomach, powerful acids destroy most of the bacteria.

These cilia explain the runny-nose-on-a-cold-day syndrome. The cold air partially paralyses the cilia, and causes an overproduction of the mucus. The mucus isn't swept backwards to the throat, but instead dribbles out at the front and onto your upper lip.

ANCIENT PERFUMES FROM POMPEII

Two thousand years ago, Pompeii was famous for making perfumes. In fact, it is the only known ancient site in the Mediterranean that includes a perfumery — the Perfumer's House. A house nearby, the House of the Vettii, has a fresco that shows a woman smelling her wrist, and angels grinding flowers and mixing extracts.

Anna Maria Ciarallo, a biologist and botanist, has used her skills to recreate the garden of the Perfumer's House. She looked at charred seeds dating back to the eruption of Mount Vesuvius that destroyed Pompeii in AD 79, and at local archaeological diggings. She also examined residues in 2000-year-old perfume bottles. Her recreated garden has roses, violets, lilies, myrtle — and seven olive trees grown for their oil. They will be used to make perfumes that will be as similar as possible to the ancient perfumes of Pompeii.

References

Rachel S. Herz, "The ascent of scent", *Scientific American*, November 1999 — www.sciam.com.

Rachel S. Herz, "Scents of time", *The Sciences*, July–August 2000, pp 34–39.

Rachel S. Herz, "Verbal coding in olfactory versus nonolfactory cognition", *Memory & Cognition*, Vol. 28, No. 6, September 2000, p 957.

Bruce Johnston, "Aroma of ancients wafts into our airspace", *The Sydney Morning Herald*, 31 March 2001, p 25.

Robert Kunzig, "A fragrant revolution", *Discover*, February 2000, pp 22–23, 88.

Stephanie Pain, "The perfume hunters", *New Scientist*, No. 2287, 21 April 2001, pp 34–37.

Neil Pendock, "Memories light the corners of my wine", *Sunday Times*, 30 May 1999.

John Travis, "Disabled genes dull sense of smell", *Science News*, Vol. 157, 6 May 2000, p 298.

Edmund White, *Marcel Proust: A Penguin Life*, Viking Penguin, New York, 1998.

Bathroom QUEUES BLUES

Why do women get the blues
When they're waiting for the loos?

Whenever you go to a sports event, concert or play, you might notice something curious happening during the interval. Everybody rushes for the toilets, but the men always make it back to their seats long before the women. It's easy to see why: the queues at the women's toilets are very long.

We've had bathroom technology for about 5000 years. So why is it so, that when a girl's gotta go, she gets the bathroom queues blues? (...Might make a good blues song ... Cain't you hear that guitar twang? ...)

EMBARRASSMENT

There's usually some embarrassment associated with natural bodily functions such as washing and eliminating various liquids and solids. I think that this is a little strange. After all, they *are* "natural bodily functions", and we all do them.

This embarrassment has given us many words for the place where we do go to relieve ourselves: bathroom, lavatory (or lav), toilet, water closet (WC), amenity, ablutions centre, throne, loo, toilet, powder room, little girls' (or boys') room, thunderbox, privy, and even that strange Australian word, dunny. People will ask to "be excused", or where they can "powder their nose" or "wash their hands". In the play *Who's Afraid of Virginia Wolfe*, the inebriated husband gets straight to the point, and invites the guests to use the "*euphemism*".

In 1530, Desiderius Erasmus of Rotterdam wrote one of the first known books of etiquette. He cautioned, "*it is impolite to greet someone who is urinating or defecating*". He also gave advice on farts: "*let a cough hide the explosive sound ... Follow the law: replace farts with coughs ...*".

Sometimes, people were more than a little impolite. In 1589, the British Royal Palace had to post a stern warning: "*Let no one, whoever he may be, before, at, or after meals, early or late, foul the staircases, corridors, or closets with urine or other filth.*"

MISCHIEVOUS BEHAVIOUR

The royalty have always been a law unto themselves — but sometimes they won't even follow their own rules. In 1665, King Charles II and his court received a very low rating on the Nice Guest Scale, after they spent the summer in Oxford. The Oxford antiquarian Anthony à Wood wrote in his diary: "*Though they were neat and gay in their apparrell, yet they were very nasty and beastly, leaving at their departure their excrements in every corner, in chimneys, studies, coleholes, cellars.*"

One and a half centuries later Lord Byron was accused of similar behaviour. He was thrown out of a London hotel because he thought "*the hall to be a less inclement place than an uncovered yard*".

PRIVATE TOILETS

A hole in the ground is at the bottom end of toilet technology.

Ten thousand years ago, the inhabitants of the Orkney Islands (off the coast of Scotland) had gone medium-tech.

CLEANING AND SOCIETY

Many societies have hygiene practices that have great symbolism. They range from bathing in the holy Ganges River in India, to the washing of the feet of newly ordained Catholic priests. This rich symbolism gives many different meanings to the concept of "clean".

For example, one rather slanderous saying about cleanliness claims that "*the Frenchman washes his hands before urinating; the Englishman washes his hands afterwards — which could be anywhere from seconds to weeks later*".

We talk about a criminal suspect being "clean" when their innocence has been established.

In Western society, "The Great Unwashed" is any group in society that you don't like. This term came about, according to Reginald Reynolds in his book *Cleanliness and Godliness*, because the British upper classes had a daily bath, which the poorer lower classes could not afford.

But the upper class shouldn't have been so uppity. Reynolds says that the British learnt to bathe regularly only recently — from the Hindus in India. He also claims that it took the British upper class several generations to make the daily bath a regular habit.

They built drains that ran from inside their stone huts to nearby streams, allowing them to relieve themselves while staying inside. But it also meant they polluted the streams.

Five thousand years ago, the homes of the Hindus had toilet facilities with plumbing. Archaeologists have found terracotta pipes fitted to baths in the Indus River Valley of Pakistan. These pipes even had taps to control the flow of water.

Four thousand years ago, there were very sophisticated bathrooms in the Minoan Palace at Knossos (in what is now Crete, a Greek island). They had hot and cold running water, and sewerage systems that used water to flush away urine and faeces. Vertical stone pipes carried water to and from the bath tubs. This palace also has what seems to be the very first flush toilet in history — a toilet with an overhead water reservoir.

Three and a half thousand years ago, aristocrats in Egypt had bathrooms with copper pipes carrying hot and cold water. But the Egyptian priests missed out on the hot water — back then, religious law stipulated that the priests have four cold baths each day.

Two thousand years ago, Rome introduced similar hygiene facilities. Around the same time, China had

WASH BEFORE THE BATH

In Japan, water shortages have long been a way of life. One way to save water is to have a few communal baths, rather than many single private baths. Another way is to preserve the bath water, rather than throw it out. But this means keeping the water clean.

So in Japan, you first wash and clean yourself using only a little water — usually while sitting on a wooden stool. Only when you are thoroughly clean will you enter the large hot communal soaking bath — the "furo" — and soak quietly. In fact, it is considered extremely offensive in Japan to bring soap into the furo.

In April 2000, two of the public bathhouses in Otaru in Japan banned all foreigners. This was after a bunch of drunken Russian sailors *"grabbed their vodka bottles, soaped up and jumped into the steaming waters of the huge tiled baths"*.

On one of my family's Outback trips, we stopped overnight at an oasis. Its large, clean pool was the only water for 300 km in any direction. To our amazement, some foreign tourists jumped in and started soaping themselves and shampooing their hair. The next day, we all had to fill our drinking water containers with soapy water.

SEWAGE AND SEWERAGE

Sewer, sewage and sewerage are all quite different.

Sewage is the stuff that gets flushed down your toilet bowl.

It then travels through the *sewer* pipes of the *sewerage* system.

Unfortunately for people learning the complicated English language, "sewer" can have more than one meaning and pronunciation. Consider this sentence: *"A seamstress and a sewer fell down into a sewer line."*

flushing toilets with running water, a stone seat and a comfortable armrest.

But these were just brief peaks of Bathroom Civilisation. In each case, after a few hundred years the society declined and people went back to using holes in the ground, or chamber pots if they lived in a town.

Chamber pots were a very important toilet accessory for some 22 centuries. They sometimes had a picture of an unpopular person in the target area. For a while, Napoleon's face was a favourite in English chamber pots.

The chamber pot led to the term "loo" for the toilet. The chamber pots were often emptied into the streets at night. But to give pedestrians a chance, those above would cry out, "*Gardyloo*!", a corruption of "*Gardez l'eau*!", which is French for "*Look out for water*!".

In medieval Europe, the toilet facilities in most of the walled castles were very "low-maintenance". The toilets were often high in the turrets and emptied directly into the castle moat. Think about that the next time you see a Hollywood hero swimming across the moat to rescue the damsel-in-distress in the castle.

PUBLIC TOILETS

Private toilets have been around for a long time. But there was a delay before public toilets were generally available.

The Cretans had public toilets in Knossos by 1700 BC. And not only did Moses give the Ten Commandments to the Jews — he also gave them very specific rules about keeping the body clean. David and Solomon built public waterworks in Palestine between 1000 and 930 BC.

The Romans were very proud of their magnificent underground sewers. They called their "cloaca maxima" ("largest sewer") the "*signifier of civilisation par excellence*".

By the 1st century AD, the Romans had the first public toilet facility for street use — the public urinal or "pissoir". Rather unfairly, only men could use these public facilities. The Emperor Vespasian installed them as part of his public works program in Rome. He cleverly paid for the urinals by selling the urine to the cloth dyers, who used it in the dyeing process.

Much earlier, by the second century BC, the Romans had built magnificent public bath/sports facilities. The mammoth complex at the Baths of Caracalla had everything. The bathrooms had separate tubs with cold, warm and hot water, and dedicated sauna rooms. There were rooms for scraping or oiling your body, rooms dedicated to hair care, and manicure rooms. You could pump iron in the well-equipped gymnasium. You could spend money in a restaurant where slaves served food and wine, or in shops on perfumes and cosmetics. The complex even had features that modern gyms/health centres don't yet have — such as a library, a lecture hall and an art gallery. This huge complex could handle 2500 customers at one time — but men only. Women had similar but smaller facilities. But the Roman Empire declined, and this massive spa complex had closed by the fifth century AD.

Sometimes, the public toilet came to you. Dr Johnson describes how in small Scottish towns in the 18th century, vendors roamed the streets "selling" their personal public toilet. They carried a large cloak and a bucket. After you paid them, you used the bucket while they protected you from public view with the cloak. This service was available in Asia Minor and Eastern Europe up until the 1920s.

FRENCH ROYAL TOILETS

The Royal Palace at Versailles was built in the 17th century. Surprisingly, it had no plumbed water for the toilets or the bathrooms — even though it housed 4000 attendants, 1000 members of the nobility and the French Royal Family. However, it did have over 260 toilet boxes scattered around the Palace for the convenience of the Royal Family.

SITTING OR HOVERING ON A PUBLIC TOILET

In one British survey, about 96% of women said that while they sit on the toilet at home, they would never "sit" on a public toilet. Instead they "hover", because the toilet seat is usually wet. This uncomfortable position has the backs of the knees locked against the front of the toilet seat, and the thighs "hovering" above the toilet seat.

Unfortunately, you need considerable muscle tension to hold this position. When this muscle tension is combined with the psychological pressure of being-in-a-hurry, you can get mis-aiming of the urine. This splatters urine over the seat, the bowl and even the floor.

The next user is horrified by this mess, and will hover at an even higher angle to avoid any contact with the seat. This produces more inaccuracy, and creates more mess — and so the vicious cycle is established.

It took a lot to embarrass the French Royal Family — they even carried out official business while seated on the toilet. Alexander Kira, Professor Emeritus of Architecture at Cornell University, wrote about this in his ground-breaking and controversial work, *The Bathroom*: "*Kings, princes and even generals treated it as a throne at which audiences could be granted. Lord Portland, when Ambassador to the Court of Louis XIV, was deemed highly honoured to be so received, and it was from this throne that Louis announced ex cathedra his coming marriage to Mme de Maintenon.*" It was reported that Lyndon B. Johnson had a similar habit while he was President of the United States.

FLUSH TOILET — THOMAS CRAPPER

A very important part of a modern toilet system is a mechanism to keep away the smells of previous sewage. Unfortunately, in Sir John Harington's flush toilet, you could see (and smell) the cesspool.

In 1775, Alexander Cumming, a British watchmaker and mathematician, patented his design for a toilet with a "stink trap". This was a simple water trap that would *"consistently retain a quantity of water to cut off all communication of smell from below"*. It successfully separated the toilet bowl from the sewerage system, and kept the "stinks" out. The design of the modern flush toilet had arrived.

In 1886, Thomas Crapper (1836–1910) came up with two improvements to the flush toilet. Firstly, he lifted the holding tank for water even higher. This meant that the falling water not only diluted the contents of the bowl, but also pushed them out. Prior to this, many flush toilets had a valve, which had to be opened to release the contents of the bowl into the sewerage system. Secondly, it had a full U-bend, to hold lots of water in the water trap. He called it the "disconnecting trap". Soon, T. Crapper & Company were installing automatic flushing urinals in Sandringham Castle.

Even though Crapper did not invent the flush toilet, he was quite a clever promoter. He even showed "sanitaryware" in his shop windows for anybody to see. This was quite shocking at the time, and some ladies supposedly fainted at the sight.

Those same ladies must have turned over in their graves in late 2000, when a stained glass window featuring a toilet was installed in St Lawrence's Church near Doncaster, England, to honour the achievements

THE IMPORTANCE OF PLUMBING

Before the advent of plumbing in cities, faeces and urine were supposed to be removed by "night-soil men". But these workers had to be paid. It was cheaper, though illegal, to throw waste into the street late at night, when nobody was looking.

As towns and cities grew bigger, lack of sanitation caused massive epidemics of infectious diseases. In the 1830s, an outbreak of cholera devastated London and 30 000 people died. The city authorities eventually saw the value of public sanitation. They slowly introduced plumbing into private homes, workplaces and public areas.

of Thomas Crapper. The church officials thought that a traditional white toilet *"might stand out too much in the window and become the focus of attention instead of Christ in Majesty, so it was portrayed in a dark silhouette"*.

There is still controversy about whether the term "crap" comes from his surname.

The "Yes" argument says that in World War I, many American soldiers stayed in England and saw flush toilets for the first time. Most of these toilets carried his company's name. "Crapper" was shortened to "crap", and that's how the name arose.

The "No" argument says that the term "crap" was already in common use by 1846, when Thomas Crapper was nine years old. It might have come from the Dutch "krappe" or the Low German "krape", meaning "a vile and inedible fish", or from the 13th century Anglo-Saxon word "crappe", meaning "chaff, or any other waste material".

However, it does seem that the American soldiers did bring the name "crapper" back to the USA in its meaning as "toilet".

And there is even a small minority who claim that Thomas Crapper never existed. But in Westminster Cathedral, amidst the tombstones of the Great and Famous, there are manhole covers that read "T. Crapper & Co. Sanitary Engineers". And his biographer, Wallace Reyburn, who wrote *Flushed with Pride*, was convinced that he existed.

By the 1840s, there were public street urinals in Paris. Again, these were for men only. Germany began to lay underground sewers in Hamburg in 1843.

George Jennings installed effective and practical public sanitation at the Great Crystal Palace Exhibition of 1851, in London. Over 800 000 men and women used his public toilets. His company also introduced the oval toilet seat.

But afterwards, when he tried to install similar public toilets around London, he was cruelly rebuffed. His proposal was "*declined by Gentlemen (influenced by English delicacy of feeling) who preferred that the daughters and wives of Englishmen should encounter at every corner, sights so disgusting to every sense, and the general public suffer pain and often permanent injury rather than permit the construction of that shelter and privacy now common in every other city in the world*".

Even so, the Sanitation Revolution had begun, and the British would soon develop massive water and sewerage systems. In 1865, Jennings installed the first public flush toilets in London outside the Royal Exchange. The fee to use this toilet was one penny — this is where we get the saying "to spend a penny". This fee stayed the same for over a century — until 1971, when decimal currency was introduced.

Women finally got some equality in Paris by the 1860s, when the "pavillons pour dames" were introduced. These public toilets were fully enclosed.

WASH YOUR HANDS . . .

In many surveys, only 60% of the users of public toilets admit to washing their hands. In public, there's a lot of peer group pressure to wash your hands. It makes you wonder how many people wash their hands in the privacy of their own homes . . .

ELIMINATION — DEFECATION

The main "elimination" actions that we do in toilets are defecation and urination.

Defecation comes from the Latin "defaecare", which literally means "to cleanse from the dregs". Much of the world squats to defecate, but in Western countries sitting is much more common. During a typical defecation, the pressure in the tail-end of the colon usually reaches between 2.5 and 6.5 times atmospheric pressure.

The faeces of an average Western adult weigh about 150 grams per "dump", depending on the diet and the state of general health. Your faeces are made up of residues of the food that you eat, dead and living bacteria and various bodily liquids. They're about 65–75% water, about 15% ash, about 15% soluble chemicals and about 5% nitrogen.

ELIMINATION — URINATION

The average adult urinates about 1.5 litres per day.

When you turn on a tap, the water usually comes out in a steady stream. But urine does not. Alexander Kira writes: "*urine passing through the slit-like urethral opening is emitted in the form of a thin sheet that twists and spirals for approximately 100–150 mm and then disintegrates into a centrifugal spray*". The spacing of the twists in the moving stream of urine depends on the exit speed. At low speed, each twist is about 10 mm long, while the twists in a high-speed urine flow are about 50 mm long. The average twist is about 25 mm long.

After a short distance, the urine "*disintegrates into a centrifugal spray*". This spreading outwards of the urine stream causes the inevitable splatter. This splatter is obvious when one urinates from a standing position.

It's easy to avoid airborne "splatter". Position yourself so that the urine stream travels only a short distance before it hits the target, ie urinal or toilet bowl. A nice extra touch would be for designers to shape the target so that there is no splash-back.

MEN AND WOMEN

The psychiatrist Sigmund Freud wrote: "*Anatomy is destiny.*" This becomes personally relevant if you're destined to stand in the queues to the women's toilets at any public event.

When urinating in public toilets, men spend less than two minutes between entering and leaving. The actual act of urination takes about 40 seconds. The rest of the time is spent adjusting their clothing, washing and drying their hands and gawking at themselves in the mirror. Men can urinate equally well from a sitting or standing position — if you ignore the inevitable splatter of the standing position.

But women usually urinate from a sitting or squatting position. Women generally have to remove or partially remove clothing — unlike men, who simply adjust their clothing. Women sometimes have to carry out other acts of hygiene, such as using tampons or menstrual pads, or applying contraceptive devices. And of course there is the occasional need to touch up the lippy. These actions all add extra time.

On average, women take about 2.3 times as long as the blokes in the bathroom.

The good doctor wrote: **"Anatomy is destiny."** *This observation is hugely relevant if you are a woman standing in a queue at any public toilet during a major event.*

There's another problem carrying over from the days when all architects were men. Even today both the ladies' and the gents' have the same number of cubicles, BUT the men also have urinals. Some would argue this is another unfair advantage in the queues debate.

DOUBLE WHAMMY

But women are further disadvantaged by a hangover from the days when all architects were men. The men's and women's toilets generally have the same number of sit-down toilets, or cubicles. But the men get urinals as well.

So the iniquitous situation at public events is that the women have half as many facilities to use, but take over twice as long. It's inevitable that the queues at the women's toilets will be enormous.

FLUSH TOILET — QUEEN ELIZABETH

Sir John Harington (1561–1612) was the godson of Queen Elizabeth I. Besides being a courtier and author, he also built a flushing toilet for his godmother.

His father's first marriage was to an illegitimate daughter of King Henry VIII. But his father's second marriage was to a commoner, Sir John's mother. So Sir John was related to the Queen by marriage, not by blood.

Sir John soon got into trouble. Queen Elizabeth had tried unsuccessfully several times to find a husband from the noble houses of Europe. Sir John translated from the Italian the story of Gioconda, which was the naughtiest part of Ariosto's rather risqué poem, *Orlando Furioso*. It may or may not have been a coincidence, but his translation seemed awfully close to what was happening in his godmother's delicate marriage negotiations. He insensitively distributed the translation around the Royal Court. She gave him the rather strange punishment of banishing him from her Royal Court until he had translated all 40 000 lines of this epic poem.

In 1589, Sir John designed and built a flush toilet for his house in Kelston, near Bath. In 1596, he installed a flush toilet for his godmother at Richmond Palace. Perhaps he wanted to get into her favour again.

His version of the flush toilet had three of the features of the modern flush toilet. First, it had a holding tank for fresh water high above the toilet. This gave enough pressure to successfully flush the

MATHS TO THE RESCUE

Over the centuries, we have increased our skills in the fields of Anatomy, Physiology and Architecture. But we still have the problem of long lines at the women's toilets. Around 1900, the Land of Mathematics gave us an essential tool to tackle the problem: Queuing Theory.

A century later we are able to put it all together. Today, we know why bathroom (and other) queues happen, and how to fix the problem.

TOXIC WASTES

Your faeces are full of germs. If these germs got into your gut or blood supply, they could make you very sick indeed. But urine is not toxic. Indeed, if you have bacteria in your urine, this is a disease situation and is called a Urinary Tract Infection (UTI).

wastes away. Second, it had a tap to let water into the tank. This meant that you didn't waste water by letting it wash continually into the toilet. Finally, his toilet had a valve to release the sewage into a cesspool below. He proudly called it *"a privy in perfection"*.

However, it had only a crude trapdoor to stop the smells from the cesspool coming up. These smells were so bad that they stopped Queen Elizabeth from using the toilet her godson had installed for her. His toilet didn't have today's water trap to exclude the smells. Queen Elizabeth did grant some 60 patents during her reign. But she wouldn't grant a patent to her godson for his toilet — because, it is said, of *"propriety"*.

That same year (1596), Sir John again did something literary that got him into trouble. Back then, the word "jake" was slang for chamber pot. The name "Ajax" was a pun on that rude word "jake". Sir John wrote a rather "earthy" book entitled *A New Discourse on a Stale Subject, Called The Metamorphosis of Ajax*. On one level, it was about obscenity and hypocrisy. But on another level, it was about the design of his godmother's toilet — the one that she could not bring herself to use. It even wished that the readers would *"find a way to cleanse, and keep sweete, the noblest part of themselves"*.

Queen Elizabeth could not allow this insult to her dignity to go unpunished. Once again, Queen Elizabeth banished him from her Court — and his flush toilet fell into disuse.

QUEUING THEORY HISTORY

Agner Krarup Erlang published his first paper on Queuing Theory back in 1909. Queuing Theory is also called Waiting Line Theory. Erlang was a mathematician working for the Copenhagen Telephone Company. He was trying to get a more efficient match between the customers, the telephone lines and the circuits in the telephone exchange.

In his honour, traffic intensity across the world is today measured in "erlangs". An "erlang" is defined as "*the number of requests for service actually made, divided by the number of requests for service that could have been made, if the channel had been running at full capacity*".

Queuing Theory tries to deliver "the best possible" service to customers who might arrive at unpredictable intervals. Queuing Theory looks at many factors. One of these is "congestion", where customers might "congest" or choke the access. A congestion can happen anywhere — in the sky where planes wait for a landing slot, on freeway entry and exit ramps, or in your local shop.

Queuing Theory received a huge boost during World War II in Britain. The Royal Air Force had only a few

Agner Krarup Erlang studied queuing and published a paper on it in 1909. Today, traffic intensity around the world is measured in "erlangs".

WOMEN AT WAR

The USS *San Antonio* is one of the first warships built with women in mind. This US$800 million marine troop carrier will carry a crew of 500, as well as 700 marines. About 40% of the crew are expected to be women.

So the female toilets (called "heads" on a ship) won't have urinals. Women use more toilet paper, so the designers have allowed storage room for seven rolls, not two. The female heads will also have extra ventilation (for hair spray) and more power outlets and mirrors (for hair care and makeup). The ship's barber will now have a sink, because many women need their hair wet before a cut. And there will be a change to the industrial-grade washing machines on board — a "gentle" cycle that won't damage women's delicate underwear.

radar units to find hundreds of incoming enemy aircraft. They needed a more efficient use of their scarce resources, so they could better coordinate Early Warning and Defence.

This new field of Mathematics was called "Operations Research". It was eventually acknowledged in 1948 as a "proper" academic field, when the Massachusetts Institute of Technology started a course in this area.

"THOSE ALSO SERVE WHO STAND AND WAIT" . . .

Queues are a part of life. Americans waste over 37 billion hours each year waiting in line.

Emergency services (such as police, fire brigade or ambulance) have to deal with queues very carefully. The time factor is very critical.

For example, the probability of a police officer arresting an offender near the scene of a crime is highest within two minutes of the crime being reported. The probability then drops very quickly to near-zero after only 10 minutes.

A fire goes through different stages — warming up, taking a good hold, and spreading further afield. Obviously, the sooner the fire trucks arrive, the less the damage.

Consider a serious heart attack. You have a reasonable chance of surviving if you get professional medical help at the two-and-a-half-minute mark. But the probability of dying rockets to very high by the five-minute mark. So a five-minute delay is not just half as good — it's almost certainly death.

BATHS IN THE USA

In the 19th century, most toilet facilities in the USA were outdoors.

But a building boom in the 1920s changed all of this. Laws were passed that every new dwelling in an urban area must have its own private indoor bathroom. That's how the standard three-fixture (shower, toilet and bath), 5 by 7 foot (1.52 by 2.13 metres) bathroom came into existence.

QUEUING THEORY

Queuing Theory looks at three major elements.

First, there is the *queue* itself. This line gets longer or shorter, as people join or get served. In most cases, people get served or processed on a first-come-first-served basis. But sometimes the processing can be random, such as when many people rush for a single taxi on a rainy night.

Second, there is the *server*. The server can be active (like the person at the cash register in a petrol station) or passive (like the petrol pumps). With the petrol pumps, there are many servers (pumps) for many queues. But you can also have the situation of a single queue with many servers — like the single line in a bank that is processed by several tellers (if you can find a bank with tellers).

DR QUEUE

Richard Larson of the Massachusetts Institute of Technology is one of the world's leading experts on Queuing Theory or Operations Research. In fact they call him "Dr Queue".

Back in the 1970s, he came up with a computer program that did a very good job of matching "resources" to "needs". He worked out one of the best ways to scatter fire and ambulance services throughout any particular city. He's worked as a consultant for many big companies such as United Artists, Coca-Cola, the United States Department of Justice and American Airlines.

Queuing Theory can save big money. Bob Bongiorno runs the Operations Research Department for United Airlines. He says, *"There are millions of possible flight paths for flying from A to B — we came up with an algorithm that searches all those options and gives us the best one. It has saved us a huge amount in fuel costs and pilot time, and shaved minutes off our flight times."* In 1997, this computer program saved United Airlines US$60 million.

The third element is the *location* where the queue and the server meet. The location has an effect on the queue. If the location is the small waiting room of a doctor, the queue won't get very long, because later arrivals simply can't get in the door. But if the location is outdoors, such as in an amusement park, then the queue can be incredibly long.

SINGLE VS PARALLEL QUEUES

By 1917, Erlang was studying the "single queue" or "combined queue" system. You often see combined queues in bank foyers, railway stations and airports. In a combined queue, there's one queue and several servers or counters. Once you're at the head of a queue, you just go to the first available server or counter.

One great advantage of a single or combined queue is that you get "social justice" — the first person *in* is the first person *out*. You don't get the "Slip Skip" phenomenon, where you *slip* down the serving order in your queue because somebody else in another queue, who arrived after you, *skips* up past you. Tempers can flare, and "Slip Skip" can turn to "Slip Slap" — giving you "Queue Rage".

There are many different types of Queue Rage. In one case of Queue Rage in a Milwaukee supermarket, a woman cut off half the nose of another woman. The offender had taken more than the maximum of 12 items to a "12 Items or Less" queue, and had refused to change to a regular queue.

Now, remember that the single queue guarantees that the first person *in* is the first person *served*. Many studies show that fast food customers prefer the single or combined queue to multiple parallel queues, because they know that nobody can "skip" past them. Back in 1986, A. Lewin compared Wendy's (single queue) with McDonald's (multi-queue), where the waiting time in the Wendy's single queue was twice as long as in the McDonald's parallel queues. The customers loved the guarantee of "social justice" so much that they preferred the single queue — even at twice the waiting time.

Dr Stephen Juan, an anthropologist from the Faculty of Education at the University of Sydney, summarised it nicely. He said, "*Queuing is part of our social contract and a form of self-defence. We give up our right to push ahead of those weaker than us in exchange for protection against those stronger than us who want the same.*"

Human psychology is incredibly important in Queue Theory.

THE HUMAN FACTOR

William James wrote in 1891: "*a day full of waiting, of unsatisfied desire for change, will seem a small eternity*".

Most people don't like "empty time". Time stretches out when you're doing nothing and just waiting. Many studies show that one minute waiting for the bus in a queue seems to last as long as two or three minutes travelling on the bus. Clever restaurants minimise "empty time" by having customers wait in a bar or a cocktail lounge, where they can have a drink or two.

Most of us dread going into a bank in the busy period between 10 am and 2 pm. The queues can be horrendously long. But in New York, the Manhattan Savings Bank ran a live entertainment program in most of their 16 branches. Customers viewed the time waiting in the queue as an entertainment experience — not as lost time. In fact, at one bank an enterprising scoundrel actually sold admission tickets to people wanting to get inside to do their banking!

EXTRA DELAY SOLVES FRUSTRATION!

One fascinating example of the blending of Queuing Theory and human psychology occurred at Houston Airport. Each morning, between 7 and 9 am, about eight airline flights landed. The passengers on these planes complained long and loud about the incredibly lengthy delays in getting their luggage from the carousel.

QUEUING ON THE ROAD

What should you do when you're in one of the two lines of traffic that have to zipper down into one line?

Probably the most polite piece of advice I've ever come across is in a British driving manual: "*Wherever there's merging traffic, follow the rule: 'let one in and go'. This is the behaviour of fair standing queues, and it seems to be the rule on the road in the UK.*"

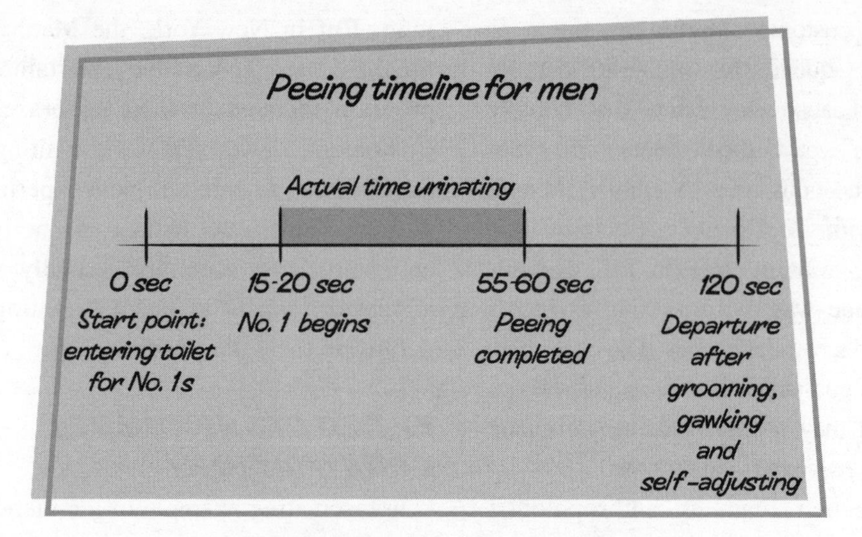

In public toilets, men spend on average less than two minutes between entering and leaving. And the actual act of urination takes about 40 seconds.

Women, on the other hand, take on average about 2.3 times longer than the chaps in the bathroom. Consider such things as sitting or squatting, removal of clothing, women's acts of personal hygiene and make-up maintenance. All fall under the obscure title of "freshening up".

Their complaints were so bitter that the airline had to hire consultants, conduct studies and even employ extra baggage handlers. The waiting time at the luggage carousel was quickly brought down to the accepted industry standard of eight minutes. But the complaints still continued with the same ferocity.

The solution became obvious only after some of the researchers took a morning flight themselves.

They realised that there was a one-minute walk from the plane to the luggage carousel, but a seven-minute wait at the carousel. While they were waiting at the carousel, they could clearly see other passengers with only carry-on hand luggage walking straight to the taxis. It turned out that most of the passengers on these early morning flights were business people. They wanted an early start to their business working day in Houston. But seven minutes of watching other less luggage-laden business people get a head start on them drove those waiting into a frenzy.

Amazingly, the cure to this problem was to add extra delays. Everything was shifted away from the taxis.

Each aircraft unloaded its passengers at the most distant possible gate in the airport. The luggage was sent to the most distant possible carousel. The total walking time to the carousel was increased from one to six minutes, and so the average waiting time at the carousel was dropped to only two minutes. Most importantly, while they were waiting, they no longer had the frustration of seeing other business people getting into taxis. Even though it took just as long to collect the luggage, the number of complaints plummeted to almost zero.

SPACE TOILET

NASA (National Aeronautics and Space Administration) spent US$23.4 million in designing and building the toilet in the Space Shuttle. It deals with what they delicately call "digestive elimination". Going to the toilet is easy down here on Earth, because gravity pulls your solid and liquid wastes down away from you. But in the micro-gravity environment in space, bodily wastes tend to just float around.

The very first "space toilet" was the simple "do-it-in-your-suit" version. Thomas Wolfe describes what used to happen in his book *The Right Stuff*. The first manned American space flight in 1961 was a 15-minute sub-orbital hop-up-and-straight-down-again, rather than a complete orbit around the Earth. The expected flight time was only a few minutes. Nobody had even thought of supplying toilet facilities. But there was a hold-up in the launch procedure. Alan B. Shepard Jr was stuck in his space suit inside the space capsule for many hours, under the harsh gaze of the TV cameras.

An astronaut with The Right Stuff would not compromise the launch by leaving the space capsule, especially while on TV — so he eventually urinated in his suit. However, Mission Control was monitoring his heart, via signals from electrodes stuck to his chest. As the warm wave of electrically conductive, electrolyte-rich urine swept up his suit and washed over the electrodes, the heart monitors in Mission Control went crazy. Soon after, Freedom 7 was launched into space. So the first American manned space flight was by an astronaut in a wet suit ...

For the next flight, the astronaut Virgil "Gus" Grissom wore a quickly-cobbled-together, enormous nappy.

The nappy then evolved into a large plastic bag with double-sided sticky tape — for sticking onto the buttocks. It had indentations in the plastic at the sticky tape end. The astronauts had to shove their fingers into the indentations, grab the faeces, and toss them down to the far end of the bag, where they would stay once the bag was rolled up. But there were a few occasions when faeces did apparently escape the bag, as it was being unstuck from the astronaut's bottom. The cameras had to be switched off to protect the sensibilities of the American public.

But the current Space Shuttle toilet, properly called the Waste Collection System (WCS), is a fully integrated, multi-function system. It is a flushing toilet — but it flushes with air, not water.

The WCS collects and processes solid and liquid biological waste from the crew members. It also accepts condensate water from the space suit used in an EMU (Extravehicular Mobility Unit), and even vents smelly gases from garbage containers overboard.

The WCS sits in its own tiny toilet in the middeck of the crew compartment. The room is 73.7 cm wide, and so is the commode. The toilet door has two privacy curtains — one at the top and one at the bottom.

The astronauts are held on the commode/toilet by a bar across their thighs. As you know, NASA always has backups for everything.

Here, the backup system to the bar is a set of four Velcro thigh straps. And of course, there are foot restraints as well.

All the gases that go into the commode are drawn through an odour and bacteria filter, where they mix again with clean cabin air. This filter can be replaced easily in orbit.

The urinal works for both males and females. It's basically just a funnel attached to a flexible hose. It can be used when you are sitting on the commode or standing. The urine is drawn along by flowing air. The mixture of liquid and air from the urinal is carried to a rotating chamber. Centrifugal force throws the liquid to the outside, from where it travels to the waste water tank. So yes, the urine does hit the spinning circular air-conditioning object . . .

The commode seat is made of a contoured, compliant, semi-soft material. Not only does it locate the user in the right position, but it also makes a good air seal with the buttocks of the astronaut. The hole for solid wastes is about 10 cm across.

When the toilet is in use, it has a continuous flow of cabin air whistling in through holes under the seat. This airflow of 850 litres per minute is essential to push the faeces toward the bottom of the toilet, because there is no gravity to make them drop. The airflow draws faeces from the commode into a porous bag. The bag is made of a multi-layer material that repels water. The bag traps solid wastes (including toilet tissues), but lets the air through. It also stops any free liquids or bacteria from leaving.

Once astronauts have finished using the WCS, they open a valve and expose the contents to the vacuum of space. All solid wastes in the bag get instantly freeze-dried and de-odourised. The bag of waste doesn't fly off into space because it's firmly attached. The wastes are then brought back to Earth for analysis and disposal.

One of the early surprises of space travel was that the solid wastes are very rich in calcium. The micro-gravity environment of space triggers the astronauts into an immediate osteoporosis, and their bones lose lots of calcium, which turns up in their faeces.

The liquid in the waste water tank is dumped overboard from time to time. One of the astronauts has said that *"there is nothing as beautiful as a urine dump at sunset"*.

DELAY CAN BE GOOD . . .

But sometimes the exact opposite of the Houston Airport solution is what you need. Sometimes, we humans actually prefer to wait in line, instead of being served immediately!

For example, at some theme parks, the queues are deliberately snaked past the rides. This builds up a sense of expectancy. A few studies show that a sense of anticipation is necessary to enjoy a ride. On a quiet day when the queues are very short, the customers don't enjoy their ride as much as when they had to wait for 20 minutes. To keep the people in line entertained, the queue will usually snake back and forth past the people already enjoying the ride. One theme park spokesperson said, "*Some people are literally screaming with excitement by the time they finally get to the ride.*"

Theme parks usually prefer to have a single, enormously long queue for each ride. The main reason is the "social justice" factor — nobody can unfairly skip ahead of you in the queue. The long queue also tells the world that the ride is very popular.

TOILET TRAINING THE OUTBACK WAY

In an urbanised society with flush toilets, there are very few good role models for toddlers to copy in toilet training. All they see is that people will excuse themselves from the room, a little later there may or may not be a quiet flushing noise, and then they come back into the room. But when our little family spent 18 months travelling through the Outback of Australia, we adults provided a very good role model.

When we were on the road, going to the toilet was a big deal. It required time and planning. First, we had to find soil that looked soft enough to easily dig a deep hole in, but hard enough not to get blown away by the wind. We parked the 4WD truck. Then we had to break out the shovel, the toilet paper, the box of matches, the shower head and hose (which we put into the bayonet — push and twist — fitting on the side of the truck), the towel and some soap.

The next stage was to dig the hole, perform the elimination actions as required, and use toilet paper. It was important to set fire to the toilet paper in the hole, so that if the hole did get uncovered by an animal the toilet paper wouldn't blow across the countryside. Then we would fill in the hole, go back to the truck, switch on the water pump and wash our hands with the soap, dry our hands, and finally put back everything that we had taken out.

This was a pretty involved process. Before long, our 18-month-old son was copying us, digging with his little plastic spade ...

SINGLE vs PARALLEL QUEUES REVISITED

Seventy years after Erlang wrote his first paper, three scientists re-examined queues. They found that in certain cases a parallel queue could have advantages over a combined or single queue.

One of the scientists was Richard C. Larson, from the Massachusetts Institute of Technology, who wrote about "Social justice and the psychology of queuing". The other two scientists, Michael Rothkopf from the University of California at Berkeley and Paul Rech from San Francisco State University, wrote about how "Combining queues is not always beneficial". These two papers appeared in the November–December 1987 issue of the journal *Operations Research.*

First of all, a combined queue "looks" longer than a bunch of short queues. This immense single queue can fill customers with dread. They could easily turn away and take their business elsewhere.

Secondly, a combined queue can dehumanise the customer. You feel like just another anonymous number in the system. You can't even choose to go to your favourite bank teller. But in a parallel queue, you can choose your teller.

Thirdly, a combined queue means that all of the servers have to be identical. You can't have specialised servers — such as a bank teller who is very skilled at processing international money orders. Each serving station has to provide all possible services — and this can be expensive. After all, it's very frustrating for you to arrive at a server, only to find that you can't get the service that you waited for.

Finally, in a single-queue system, even the customers who can be processed quickly are forced to wait in the one long queue. This makes the queue longer than it need be. If you could feed these customers to a "quick-processing" queue you'd reduce the overall queue length, because the fast-processed customers would all leave the premises more quickly.

TOILET WATER — TOILET LID UP OR DOWN?

There are many small but consistent differences in male-female behaviour in Western society. One is that males tend to leave the toilet seat and lid up. There's an easy way to fix this.

Just ask the offending male, *"Do you like to brush your teeth with toilet water?"*

If they flush the toilet with the seat and the lid up, the turbulent water liberates a fine mist of toilet water all over your bathroom. Some of it lands on your toothbrush. However, most people prefer to get the "ring of confidence" from their toothpaste.

It's easy to fix. Just lower the lid (and the seat of course) before you push the button.

THE FASTEST QUEUE AT THE SUPERMARKET

But what can you do if you get stuck behind a person who needs a lot of processing time? Consider the situation of backpackers, with very limited English-language skills, entering a bank with a 5-kg bag of small denomination coins from 27 different countries that they have just visited. They want to change these coins for the five different currencies of the next countries they'll be visiting.

The solution is easy — just jump to another lane.

Part of queuing paranoia at the supermarket is the sure knowledge that you're almost certainly going to be beaten by one of the queues next to you. This paranoia is real. Assume that all the queues are the same length. Say that you've got 12 queues. Your chance of being processed first is one in 12. In other words, 11 out of every 12 times you go shopping, one of the other queues will beat you.

But this is how to choose the quickest queue at the supermarket: choose a queue that is on the outside. The queue that is surrounded by other queues can be jumped onto from each side. But a queue on the outside can be jumped onto only from one side.

But if that queue stops moving, then that's your cue to move too ...

QUEUING THEORY AND BANKS

Banks like ATMs (ATM = Automatic Teller Machine = Money Wall = Hole in the Wall). If they have an ATM, they don't have to pay the wages of a teller. Banks want to save even more

Let's hope that the future holds better prospects for us all AND a policy to lose the bathroom queues blues.

money, so they install as few ATMs as possible. But the customers want the queue at the ATM to be as short as possible, so they want plenty of ATMs.

The banks follow a general rule of thumb about waiting time. They hope that 80% of the time the customer will wait for less than five minutes.

Dr Sam Savage is the Director of the Industrial Affiliation Program at the Department of Engineering Economic Systems and Operations Research at Stanford University. He has researched this problem using Queuing Theory. His calculations tell us that if an ATM is in use 80% of the time, there will be an average of 3.2 people waiting in the line. If the ATM usage goes up to 90%, the bank will be happier — but the customers will be very unhappy with an average queue length of 8.1 people.

And if the ATM is used 100% of the time, then the number of people in the queue ahead of you will be infinite!

QUEUING THEORY AND BUSES

One famous example of Queuing Theory is the Bus Paradox. You arrive at an airport and drag all your luggage to the bus stop, where you read that buses arrive every 10 minutes. So how long would you expect to wait for a bus? On average, you've probably arrived between two buses, so you'll have to wait for five minutes, right?

Nope. Queuing Theory tells us that buses get held up by random causes, such as traffic jams, or people with too much luggage trying to haul it all onto the bus. So the interval between buses changes from a regular 10 minutes to random intervals. Sometimes two buses will arrive at the same time, and if you miss them you'll have to wait 20 minutes.

According to the maths of Queuing Theory, the average time that you'll have to wait will be the advertised interval between buses — 10 minutes.

QUEUED BUSES COME IN THREES

For over a century, enraged commuters have had a problem. They wait for ages for a single bus, and then suddenly three will come at once.

Professor Tony Wren is a timetabling expert from Leeds University in the north of England. He says that there are five stages: "*... first, the actual blockage; next, passenger queues build up; third, people take longer to board... and the next bus catches up; fourth, hardly anyone is left for it and even fewer for the third; and finally... the cluster of buses remains a trio*". It is possible to reduce the Triple Bus Problem by holding the second or third buses back to increase the gaps. Unfortunately, this can throw out the timetable even further.

QUEUING THEORY AND TOILETS

Queuing Theory is being given further consideration in banks — and in women's toilets. It's about time. In fact, in some sports events or theatres, the managers have extended the interval to give the queuing women enough toilet time.

Queuing Theory tells us that if women spend twice as long in the toilets, their queue will be at least four times longer. Luckily, Queuing Theory also gives us a solution, which many women will be busting to hear. All we have to do is give women twice as many toilets, and the queues for the men's and women's toilets will be the same length. True equality at last. Indeed, at the 2000 Summer Olympics in Sydney, the main Olympic Stadium was built with three times more toilets for the women than for the men.

So how come it took the human race 4000 years to work that out?

References

Alexander Kira, *The Bathroom*, Viking Press, New York, 1976, pp 3–26, 113–117, 146–157, 193–199, 223–237.

Richard C. Larson, "Perspectives on queues: Social justice and the psychology of queuing", *Operations Research*, Vol. 35, No. 6, November–December 1987, pp 895–909.

Martin A. Leibowitz, "Queues", *Scientific American*, August 1968, pp 96–103.

Robert Matthews, "Hurry up and wait", *New Scientist*, No. 2091, 19 July 1997, pp 24–27.

Robert Matthews, "Ladies in waiting", *New Scientist*, No. 2249, 29 July 2000, pp 38–39.

Rosie Shaw, "Queue for the loo", *New Scientist*, No. 2252, 19 August 2000, pp 52–53.

Lawrence Wright, *Clean and Decent: The Fascinating History of the Bathroom and the Water Closet — And of Sundry Habits, Fashions and Accessories of the Toilet — Principally in Great Britain, France and America*, Routledge and Kegan Paul, London, 1960, pp 75–78, 200–210.

Farts — BEGONE WITH THE WIND

What's in a fart, and how do they start?

Farts are such an embarrassing concept that the English language doesn't even have a verb for them. Instead we simply use the noun as a verb. However, thanks to my medical training, I can handle such concepts without embarrassment.

FARTS IN THE OPERATING THEATRE

So I felt quite at ease when a surgical operating theatre nurse rang me with an important medical question about farts.

Her busy day in the operating theatre suddenly developed a quiet period, after one case was cancelled. There would be a 20-minute delay. It would take her 20 minutes just to get out of all of her sterile theatre clothing, scrub up and then put on new sterile theatre clothing. So she decided to stay in her sterile theatre gear. As she stood there alone in the empty operating theatre, she suddenly had the desire to "pass wind". There wouldn't be anyone else in theatre for at least another 10 minutes, so she did.

She immediately began to wonder if she had done a bad thing, and had inadvertently contaminated the sterile environment of the theatre with germs. Would the next patient be at risk of an infection? So she rang my Triple J Science Talkback show, and asked me.

Easy questions often have difficult answers. This would be the case with farts, as I found out when I went looking in the literature. The closest answer I came across was the case of an anaesthesiologist who probably infected patients with Group A *streptococcus* by farting.

LE PETOMANE

Each day, the average human breaks wind about 7 to 14 times, and releases between 200 and 2500 ml of gas via the anus. For most of us, "dropping your guts" or "cutting the cheese" is just a social embarrassment. But Le Pétomane made a career out of well-controlled flatulence. Le Pétomane means "The Fartiste" or "The Manic Farter".

Joseph Pujol was born in Marseilles in 1857. He soon discovered that he could control the muscles of his abdomen like a bellows. When he synchronised this with careful control of the muscles of his anus, he found he could easily move 2 litres of water in and out of his colon — purely under voluntary control. He soon progressed to moving air in and out.

He started off doing tricks for his friends. Of course, he would give his colon a thorough wash-out before any performance, so there were never any embarrassing odours. His "natural" vocal range was only four notes — *do, mi, sol* and *do* again.

His first professional performance in 1887 met with mixed reviews, but he persisted. His career rocketed when he began performing at the Moulin Rouge music hall in Paris in 1892. He would appear onstage elegantly attired in red cape, black trousers and white cravat, with a pair of white gloves casually held in his hands.

His sound impressions included bird songs, the timid fart of a bride on her wedding night, her lusty raspberry fart one week later, and a imposing 10-second fart that sounded like the cutting of coarse cloth.

After more sonic impressions, including farting "La Marseillaise" and "Au Claire de la Lune", he would delicately retire offstage to insert a tube into his anus. Back onstage, he would further entertain the audience by using the tube to play various wind instruments, and to smoke a cigarette right down to the butt. He could even blow out candles from 40 cm away.

He was immensely popular. At one stage he was earning 20 000 francs per week, two and a half times more than the actress Sarah Bernhardt. Even the King of Belgium came incognito to see him perform. But the battles of World War I disabled two of his children, so he retired from showbusiness, and he finally died in 1945.

THE MAGNIFICENT FART

The French gastronome Jean-Anthelme Brillat-Savarin realised that "*The intestines are the home of tempest. In them is formed gas in the clouds.*"

The average fart is a wondrous event. Let me demonstrate with a little experiment. Link the fingers of your two hands tightly together to make a little cup, with the palms facing upwards. Imagine that in the cup of your hands you have some water, some floating solids and some gas. Imagine the whole system is under pressure. Now try to open one of your fingers in such a way that you release only the gas, without letting any solid or liquid squirt out.

I really doubt whether any device made by the human race could do this task — and keep on doing it for some 70 years.

This is the magnificent job that your anal sphincter does some 10 times per day, with very few failures. The proctologist Dr W.C. Bornemeier wrote in admiration of the anal sphincter: "*It apparently can tell whether its owner is alone or with someone, whether standing up or sitting down, and whether its owner has his pants on or off. No other muscle in the body is such a protector of the dignity of man, yet so ready to come to his relief.*"

THE EMBARRASSING WORD, "FART"

Farts are considered so embarrassing that we still don't have a commonly accepted polite word for them. "Flatus" is too technical and scientific and cold. The word "fart", according to Kenneth Heaton, a gastroenterologist at the Bristol Royal Infirmary, "*is even less acceptable in polite conversation than faeces*".

As a result, we have many euphemisms for farts. They include "go off", "let off", "lunch", "make a smell", "shoot a fairy", "fluff", "bum burp", "drop a bundle", "cut the cheese (or mustard)", "braff", "break wind", "drop one's lunch (or guts)", "let one rip" and "let Fluffy off the chain".

On one occasion the *New England Journal of Medicine*, one of the leading medical journals in the world, tried to address this issue, and proposed several "nice" alternatives to the word fart. The list of invented words included "exogust", "flatulate" and "boomerate". The word that won the most votes was "exmeteorate", but it was never accepted. So a poem was composed to commemorate this lack of a polite alternative to "fart":

As a matter of fact,
The state of the art
Denies us a word
That is better than fart.

FARTS AND EXPLOSIONS

There is even the occasional surgical hazard associated with farts. The fart gas can contain methane (5–13%) or hydrogen (4–74%). Both of these gases are inflammable. There have been reports of "big bangs" when surgeons were using electrocautery machines during operations on the colon.

Go on, pull my finger!

The proctologist Dr W.C. Bornemeier had a great admiration for the anal sphincter. He marvelled as its ability to protect the dignity of human beings, yet be so ready to come to our relief.

THE EMBARRASSING EVENT, "FARTING"

It is absolutely normal to pass wind some 10 or so times per day, but it seems we humans find it embarrassing.

Over two centuries ago Benjamin Franklin, that amazing American all-rounder, advised the Royal Academy of Brussels as to what field they should choose for a prize to reward scientists: "*My Prize Question therefore should be: To Discover some Drug, wholesome and not disagreeable, to be mixed with our common Food, or Sauces, that shall render the natural discharges of Wind from our Bodies not only inoffensive, but agreeable as Perfumes.*" Perhaps genetic engineering will help to solve this problem one day …

Back in the late 1500s, the then Earl of Oxford, Edward de Vere (1550–1604), deeply embarrassed himself with a badly timed fart. (Today, he is the prime contender for the authorship of Shakespeare's works after Shakespeare himself.) John Aubrey (1626–1697) wrote of this event in *Brief Lives*: "*This Earl of Oxford, making of his low obeisance to Queen Elizabeth, happened to let a fart, at which he was so abashed and ashamed that he went to travel, seven years. On his return the Queen welcomed him home, and said 'My Lord, I had forgot the fart'*."

But the very fact that we generate gas from our bowels and pass wind means that the 400-plus friendly and essential bacteria that live in our large intestine are doing very well. Being embarrassed about farts is as silly as being embarrassed that we have legs.

FART RESEARCH

Indeed, farts are seen as so embarrassing that there hasn't been a huge amount of medical research on them. (This is probably also because farting isn't considered to be life-threatening.)

However, Dr Michael Levitt has been researching farts for many years. He is informally known as "The King of Gas". Back in the 1970s, he wrote in the *New England Journal of Medicine* of the rise of what he hoped would be a new medical speciality: flatology. He pointed out several cases where analysis of the flatus gas composition had provided valuable clues to the patient's diagnosis. He predicted that farts would become the "*rightful province of both flatologists and scatologists*".

The gastroenterologist Professor Terry Bolin has also been interested in farts for many years. He has pointed out that there is not a single medical case on record of anybody ever dying from excessive active farting, or from passive farting (eg, being in the same lift as an active farter).

In 1997, Terry Bolin and the nutritionist Rosemary Stanton released their ground-breaking book, *Wind Breaks*. They were virtually forced into writing this book, because they found they could not answer basic questions from their patients about farts. They carried out a study of flatus emissions in 120 healthy men and women, to find out what was "normal" for this group.

They found in their study that the average number of farts for women was about seven per day, but 12 for men. This was probably because men eat more food than women. More food means that the bacteria in the large intestine have more to work with.

In general, the amount of farting was related to how much fibre the subjects ate. Some high-fibre eaters had up to 30 farts per day. In general, it's good to have a high-fibre diet. But a high-fibre diet has the "disadvantage" of making you fart more frequently. On the other hand, if you cut down on fibre in your diet, you increase your risk of constipation and bowel cancer.

DO OLDER PEOPLE FART MORE?

As you get older, you suffer a generalised decrease in muscle tone.

This drop in muscle tone also happens in your large intestine. At the same time, the elasticity of your large intestine also drops.

So when you have some gas in your large intestine, it's more sensitive to being distended or stretched open. It can't stand the slightly unpleasant sensation as much as it was once able to.

Your body reacts by getting rid of the gas more frequently, which is why you fart more as you get older.

Another finding was that women tended to pass flatus mostly when they were in the bathroom. However, men were much less inhibited.

In this study, the farts of men also tended to be more "aromatic" than those of women. This was probably because men eat more spices, and also eat more compounds that contain sulphur, such as are present in meat or eggs (see **Eggstasy**).

DO EX-SMOKERS FART MORE?

Why do some people suddenly complain of excessive flatulence after they have given up smoking?

There are over 4000 chemicals in tobacco smoke. Some of them speed up the movement of food through your gut. Many smokers take advantage of this by having a cigarette while sitting on the toilet, first thing in the morning. This guarantees a speedy bowel motion.

But when these chemicals are denied to the gut there is a period of readjustment, during which time the food moves more slowly. It spends a greater time in the large intestine, and ferments — and this is why some ex-smokers suffer a temporary increase in flatulence.

HOW FARTS FORM — CHEMICAL BONDS IN FOOD

There are three main types of food: fats, proteins and carbohydrates.

Proteins are often broken down into amino acids, which are then joined back together in a different order to make a different protein.

The fat and carbohydrate molecules are often broken down to give energy. Carbohydrates are also called sugars. The common "sugar" that you buy in the shop is sucrose.

There are very many different types of carbohydrates. A very common type is based on five or six carbon atoms arranged in a ring.

If there's just one ring, it's a monosaccharide, such as glucose or fructose. If there are two separate carbon rings stuck together, it's a disaccharide, such as lactose (the main sugar in milk), sucrose or maltose. Oligosaccharides have three to six carbon rings, while polysaccharides have more than six rings (sometimes thousands). Most carbohydrates are polysaccharides (such as starch, inulin, glycogen and cellulose).

There is energy in each of the chemical bonds that hold the individual rings together. Inside each ring, there is also energy in the chemical bonds joining each pair of carbon atoms.

When you break those bonds apart, that energy is made available to your body. You use up some of that energy in joining two oxygen atoms to each carbon atom to make carbon dioxide. You get rid of the carbon dioxide by blowing it out of your mouth. But there is still a huge amount of energy left over —

and you can use that energy for other purposes.

The reason for eating fats and carbohydrates is to grind them very finely, so that the individual molecules can be sent to the liver, where they are further broken down for energy.

HOW FARTS FORM — PHYSIOLOGY OF THE GUT

Eating starts in the mouth, where your teeth cut and grind the food. Saliva (1.5 litres per day) starts the process of digestion and also helps you swallow the food.

It slithers down your oesophagus and into your stomach. A lot of grinding and mashing happens to the food in the stomach — but there's not a lot of absorption going on. Virtually only alcohol and other drugs get absorbed in the stomach, which is the first part of the small intestine.

The rest of the small intestine does the major absorption of the food you eat. Various chemicals are released from the pancreas and the gall bladder, to help you break down fat, proteins and carbohydrates.

Eventually, the big molecules of fat, proteins and carbohydrates are broken down into molecules that are small enough to cross the wall of the small intestine. These molecules enter the blood vessels in the wall and then go to the liver, where they are reprocessed into other useful chemicals and/or converted into energy.

SOUNDS LIKE A COVER-UP

Aside from the smell of farting, the sound is the other embarrassing aspect of "letting off".

If a person feels that they are about to break wind and they don't wish to be embarrassed by it, they can try dampening the sound. One term used to describe this technique is "cushion creeping". A cushion creeper is a person "*who discreetly manoeuvres their buttocks on a chair in order to minimise brap (sound) and thereby silence a fart*".

"*He who smelt it, dealt it*" refers to the farter's mentioning of the smell to others in an attempt to cover up the fact that they've just delivered a "silent but deadly". By making the first mention of the fart odour, you take the attention away from yourself and direct the blame towards others, who immediately begin to discuss the origin of the fart.

But there's a major flaw in this tactic, which is if you're the first person to smell it, that could indicate that you're closest to the source ...

HOW FARTS FORM — GASES IN THE GUT

In the stomach, the gas is similar to the air in the atmosphere, but the oxygen concentration drops from 21% to 15–16%. This happens because the oxygen is absorbed by the blood in the capillaries in the stomach wall.

In the average human, the gut will hold between 150 and 500 ml of gases at any given time. However, in the case of obstruction of the small intestine, doctors have seen pockets holding 3500 ml of gas. These can cause intense pain.

There are actually three main sources of the gases in your large intestine.

About two-thirds of the gases in your large intestine come from swallowing air. Each time you swallow, you simultaneously gulp down 2–3 cc of air. Even drinking 10 cc of water will add another 17 cc of air to your gut. Ironically, the more you worry about swallowing air, the more you unconsciously swallow. This can lead to a nasty vicious loop. If you fart a lot because you swallow a lot of air, this will tend to make you worry about the air you swallow — which in turn makes you swallow even more air, and fart even more again.

Eating also generates farts. The act of eating increases the muscular activity in your gut, via various nerve pathways. This increases the rate of food delivery to the friendly bacteria, which increases the gas production.

A very small amount of the gases in your large intestine has nothing to do with swallowing gas or eating food. The cells lining your gut are continually dying, and continually being replaced by new cells. The bacteria in your large intestine eat your unwanted dead cells

CATTLE AND SHEEP PASSING WIND

Back in the mid 1980s, it was breathlessly reported by the world media that cattle farted about 150 litres of methane per day. Since then, follow-up studies have looked more carefully at the wind production of cattle and sheep.

Cattle produce about 200 litres of methane per day, but it nearly all comes from the mouth, as burps. Sheep produce about 20 litres of methane per day.

Methane is a very powerful greenhouse gas, about 21 times worse for global warming than carbon dioxide. In Australia, cattle and sheep produce about 14% of our total greenhouse emissions. The CSIRO is looking at vaccinating cattle and sheep against the bacteria that make methane.

The anal sphincter manages to release a fart on average 10 times a day.

and mucins (which are part of the mucus lining of your gut), thus generating useful chemicals as well as waste gases.

About one-third of the gas in your gut comes from the action of bacteria. This is what happens in the case of beans.

BEANZ MAKE YOU FART

It was only as recently as the 1970s that scientists began to discover why beans make you fart. The trouble begins in the first part of the gut (the small intestine) where food is broken down and digested.

In most carbohydrate foods there are very long chains of simple single sugars that are joined together by chemical bonds. Normally, these chemical bonds are broken down by the digestive enzymes in the small intestine. But in beans, not all of the chemical bonds can be broken down into separate, single, simple sugars. You end up with short chains of less than 10 simple sugars. These short chains then pass down into the large intestine.

BEANS, WONDERFUL BEANS

Typical beans include haricot beans, soybeans, red kidney beans, string beans, broad beans, lentils, chickpeas and navy beans. These beans are actually the seeds of legumes. Legumes are a family of magnificent food plants.

Firstly, they make the beans that we humans can eat as food.

Secondly, legumes fix nitrogen. That means they drag nitrogen out of the atmosphere and then put it into the soil as a fertiliser. This is good for the next bunch of plants that you put into the same soil.

The human large intestine has been colonised by over 400 different species of bacteria. Some of them can break down the short sugar chains by fermentation. So the bacteria end up making short-chain fatty acids and various gases.

These short-chain fatty acids are very useful to you. Firstly, they include chemicals such as butyrate, which can help reduce cancer of the colon. Secondly, the short-chain fatty acids include other chemicals such as acetate and propionate that control how much cholesterol your liver makes. Thirdly, the short-chain fatty acids provide energy for your colon cells to do their normal metabolic activities. And finally, these fatty acids provide energy for the 400 or so different species of bacteria that live in your large intestine. These bacteria make up much of the solid part of your faeces (see **Bathroom Queues Blues**).

Along with the short-chain fatty acids, the bacteria also produce hydrogen, carbon dioxide and methane — and small quantities of smelly gases.

At the beginning of the large intestine there is still a little bit of food absorption going on. These gases can dissolve in the blood supply of the gut. They then travel to the lungs, and finally come out through the mouth, giving you bad breath.

But most of the gases don't get absorbed. Instead, they erupt out of the anus in a burst of wonderful malodorous music. So, with gases coming out of the mouth and the anus — beans get you, both coming and going.

FARTS AND PLANES

In World War II, fighter pilots were ordered not to eat beans. The gas in their gut would expand at high altitudes and cause extreme discomfort — sometimes, enough to incapacitate them.

MILK MAKES YOU FART

Eating lots of beans is one way to generate more gas. However, there are some other conditions in the human gut that can lead to increased gas production.

We all drink some kind of milk when we are small babies. The main sugar in milk is lactose. We break down the lactose with an enzyme in the small intestine called "lactase". Lactase breaks down the disaccharide lactose into two smaller monosaccharides — which can then be further broken down into individual carbon atoms.

But two-thirds of the people on the planet (eg, about 85% of Asians and Africans and 10% of Caucasians) lose this enzyme as they mature into adults. When they drink milk, the lactose gets all the way down into the large intestine. Water rushes into the large intestine to dilute the lactose, and this causes diarrhoea. The bacteria eventually ferment the lactose to generate lots of gas.

DEFARTERISE FOOD — 1

Beans are a magnificent food. They have several advantages besides their protein content. Because they are digested slowly, they deliver their load of carbohydrate into the bloodstream as a gentle surge, rather than a sudden spurt. This means that you don't get that "insulin spike" in the bloodstream that is thought to be involved in mature-onset diabetes. However, if you are thinking of changing your diet to incorporate beans, do it slowly to give your gut, and its load of many bacteria, time to adjust.

But as every good fartologist knows, the best farts come from beans. Even Pythagoras knew about this — as Richard Burton (1577–1640) put it in his *Anatomy of Melancholy*: "*That which Pythagoras said to his scholars of old, may be forever applied to melancholy men, A fabis abstinete, eat no beans.*"

However, you can soak dried beans in water. The oligosaccharides, which would otherwise make their way down into the large intestine and be turned into short-chain fatty acids and fart gas, will then leach out into the water instead. In the case of black-eyed beans and pink beans, 90% of these offending chemicals go into the water, which you can then toss out. About 40% of these chemicals leach out in the case of kidney beans and cannellini beans.

DEFARTERISE FOOD — 2

There is another, rather high-tech method to achieve the fart-free bean.

V.S. Rao and U.K. Vakil from the Bhabha Atomic Research Centre in Bombay (Mumbai), India, developed this new technique. They first soaked the seeds of the legume "green gram" in distilled water for 16 hours. They then blasted these beans with 250 000 rads of gamma-radiation from radioactive cobalt–60. This is an incredibly high dose; 10 000 rads is enough to kill a human within five minutes.

WHOOPI GOLDBERG

How did Whoopi Goldberg get her name, when her birth name was Caryn Johnson? Easy — she just farted a lot.

Back in the early 1980s, her colleagues at a San Francisco acting company gave her this nickname, in honour of the joke cushion that makes a farting sound when you sit on it. For a while she was called Whoopi Cushion. Then she decided to Frenchify and fancify it to Whoopi Couchant.

Then she changed it to Whoopi Johnson, but didn't really like it. She finally looked through all of her family names, and came up with the surname of Goldberg — hence "Whoopi Goldberg".

The simulated fart experiment

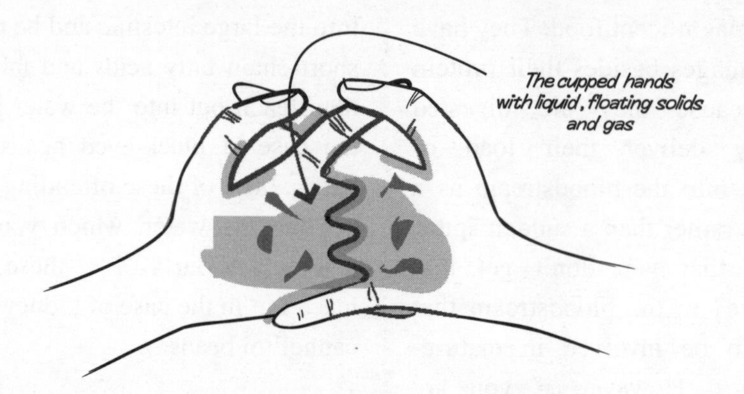

A fart is an amazing event. Imagine linking your hands and pretending they contain water, floating solids and gas AND that the whole thing is under pressure. Now try and move just one finger to release the gas only. Herein lies the brilliance of the sphincter.

The radiation did not leave the beans radioactive. Instead it weakened the chemical bonds in the sugar chains. This meant that the digestive enzymes in the human gut could break all the chemical bonds in the beans. So there were no short sugar chains for the bacteria to eat, and no gas produced.

So perhaps the burpless, fartless bean is just around the corner. *Blazing Saddles* and *Gone with the Wind* will never be the same.

DEFARTERISE FOOD — 3

If you eat tempeh with your beans, it can break down their long carbohydrate chains.

Tempeh is a compact cake of plant substance that has been completely penetrated by the fungus genus *Rhizopus*. Tempehs are very popular in Indonesia and contain as much as 40% protein. This protein comes from the invading fungus.

The word that comes after tempeh tells you what kind of plant was fermented. "Tempeh kedele" is made from soybeans ("kedele" is Indonesian for soybeans). "Tempeh bongkreg katjang" is made from peanuts, while "tempeh enthoe" comes from coconuts. Tempehs are not eaten raw, but usually deep-fried.

DEFARTERISE FOOD — 4

Cabbage and other sulphur-containing vegetables are notorious for generating farts. This is because they are rich in sulphur-containing chemicals. The heat of cooking changes these chemicals into other chemicals that can assist in the production of farts.

It turns out that the amount of these fart-generating chemicals doubles after five to seven minutes of cooking. So if you just gently stir-fry them for two to three minutes, you greatly reduce the amount of chemicals and fart gas generated.

FARTS AND LITERATURE

Geoffrey Chaucer lived from 1340 to 1400. Those times in many ways were much more ribald and open than our current Western society. His masterpiece, *The Canterbury Tales,* deals with the stories told by a group of pilgrims. They are travelling to the shrine of St Thomas à Becket, and they entertain each other on the long journey by telling stories. This is part of The Miller's Tale:

This Nicholas anon leet fle a fart,

As greet as it had been a thonder-dent,

That with the strook he was almoost yblent;

And he was redy with his iren hoot,

And Nicholas amydde the ers he smoot.

Of gooth the skyn an hande-brede aboute,

The hoote kultour brende so his toute,

And for the smert he wende for to dye.

As he were wood, for wo he gan to crye,

"Help! water! water! water! help, for goddes herte!"

For those not familiar with Middle English, the story goes roughly as follows: Nicholas and Absalom are rivals for the affection of the fair Alison (already married). At this point in the tale, Nicholas lets out a fart as great as a clap of thunder, and Absalom is almost blinded by the force of it. Absalom retaliates by striking him with a hot iron right in the arse ("*ers*"), which is so painful Nicholas thinks he's dying. He cries out for water to ease the burning — but the irony is that if he drank any, he'd only fart more!

WHAT SMELLS IN FART GAS?

Nobody is really sure why farts are so embarrassing. Some psychologists say that people wouldn't really mind farts — if they didn't smell so bad.

Most of the infrequent medical/ scientific studies have been concerned with measuring the volumes of the major gases in the flatus. These gases are oxygen, nitrogen, carbon dioxide, hydrogen and methane. Mind you, the volumes are very variable. The amount of methane generated can vary by as much as 10 million to one from one person to the next.

Other studies looked at the gases that cause the smell. But they didn't try to measure how much of each bad-smelling gas was present.

Finally, in 1998, Dr F.L. Suarez, Dr J. Springfield and Dr M.D. Levitt ("The King of Gas") from the Minneapolis Veterans Affairs Medical Center looked at this problem in a new light. In the introduction to their report in the journal *Gut*, they wrote: "*Rectal gas has been a topic of scientific and scatological interest since the beginning of recorded history. This fascination with flatus has little to do with its volume, 200–2500 ml per day, but rather its offensive odour.*" They worked out what chemicals cause the smell, and the concentrations of those chemicals. The offending chemicals all contained sulphur.

SMELLING GAS TEST PROCEDURE

They collected the flatus from 16 healthy subjects using a rectal tube. The researchers made sure that the subjects would generate gas by giving them two meals of pinto beans (200 grams the

YOU HAVE THE RIGHT TO REMAIN SILENT . . .

Thanks to "natural justice", a man who was accused of farting in a police station has been "let off".

On 13 August 2000 David Paul Grixti, of Werribee in Victoria, "poked the rear end of his body out" and "let flatulence escape" in a public space (the Werribee Police Station). Senior Police Constable Shane Andrew Binns told the court that Mr Grixti was staring at him while he farted, that he believed that the "flatulence" was intentional, and that he was "disgusted" by this fart. Mr Grixti was convicted of Offensive Behaviour, and fined $200 in early 2001.

But in September 2001, he appealed against the decision in the Victorian County Court. Judge Leslie Ross dismissed the previous conviction because farting was a "*natural occurrence that overcomes human beings, and it's usually something that can't be brought on*".

Dr Heaton (from the Bristol Royal Infirmary) stated that the word "fart" is less acceptable in polite conversation than "faeces". The ***New England Journal of Medicine*** *addressed this with some "nice" alternatives to the word "fart": "exogust", "flatulate", "boomerate" and the most popular ... "exmeteorate".*

night before, and the same at breakfast on the day of the study) and 15 grams of lactulose (on the day of the study).

The gases went from the rectal tube to a special gas-impermeable bag. In previous studies they had found that the odoriferous gases reacted with glass, rubber and some plastics, but not with polypropylene — so they transferred the flatus from the bag to a 20-ml polypropylene syringe.

They used gas-chromatographic/ mass-spectroscopic analysis to work out which particular gases were present, and in what concentrations.

Finally, they presented the flatus (in the polypropylene syringes) to the two judges to evaluate the nastiness of the odour. The judges had, in previous studies, shown that they could correctly identify a variety of different odours, and even pick threefold differences in concentrations of these sulphur-containing gases.

The three doctors' report says: "*In an odour-free environment, the judges held the syringe 3 cm from their noses and slowly ejected the gas, taking several sniffs. Odour was rated on a linear scale from 0 (no odour) to 8 (very offensive).*"

SMELLING GAS CONCLUSION

The study found that the smell of flatus comes from various sulphur-containing compounds.

The compounds in a typical flatus are hydrogen sulphide (600 million billion molecules per litre), methanethiol (120 million billion molecules per litre) and dimethyl sulphide (50 million billon molecules per litre). The judges thought that hydrogen sulphide smelt like "rotten eggs" and methanethiol like "decomposing vegetables", while dimethyl sulphide actually smelt "sweet".

"TOOT TRAPPER" TEST PROCEDURE

The scientists then showed that they had both scientific and social consciousness by testing a device that was supposed to remove the odour from farts. This would be very useful for those unlucky people who generate lots of very smelly fart gas and who have to work with people face-to-face.

The device is the commercially available "Toot Trapper". It is a polyurethane foam cushion covered with fabric. One side is covered with activated charcoal.

For the test, the eight volunteers wore "*gas-tight pantaloons fashioned of metalised nylon low-density polietilenum which were sealed to the skin at the waist and thighs with duct tape*". The pantaloons had room for the cushion inside them, and also had two access ports to allow gas sampling on either side of the cushion. A standardised test flatus gas was infused.

"TOOT TRAPPER" CONCLUSION

The Toot Trapper worked. It reduced the offensiveness of the odour of the volunteers' standardised test foecal gas. It was about 11 times better than using no cushion at all, and about five to six times better than using a cushion that had no charcoal.

MALE vs FEMALE FARTS

They also found an interesting difference between the offensiveness of the farts of men and women.

In their rather small sample, they measured that women's farts had higher concentrations of hydrogen sulphide. The judges agreed, and thought that the women's farts had a significantly worse odour. (This is the opposite of the findings of Bolin and Stanton — see *Fart Research* earlier — and just shows how little we really know about farts at this early stage.) However, men generated a greater volume of gas per passage. This compensated for the men's lower concentration, so that both men and women delivered roughly the same amount of hydrogen sulphide per passage of gas.

However, the authors of the paper did emphasise that the sample size was very small, so their results aren't especially reliable.

FARTING NURSE QUESTION

So what could I tell our farting surgical nurse with the professional and social conscience?

From all of my reading, I had not found a single thing that would help answer her real question.

Luckily, Luke Tennent happened to be listening in to my radio show that very day. He works in a microbiology laboratory in Canberra. After some discussion with a colleague, the colleague went home and got her eight-year-old son to fart onto a blood agar plate, with his pants down, from a range of 5 cm. Luke incubated the agar plate at 36°C for 24 hours, and then examined the plate. (Agar is a jelly made from algae and seaweed that holds a bacteriological culture plate together. The blood usually comes from horses and provides excellent nutrition for many different types of bacteria.) Luke called this snapshot of the close-range pattern of bacterial emission from the anus a "Fartograph".

He noticed two distinctive clusters of bacteria, and named them accordingly.

First, there was the central IBZ (Initial Blast Zone). It was made almost entirely of bacterial colonies of *E. coli* — a bacterium known to live in the colon.

Second, there was the SR (Splatter Ring). This was not a complete ring, but rather two separate arcs that, if extended, would make a continuous ring. It was made up mainly of the skin bacteria *Staphylococcus epidermidis*, and bacteria that live in our environment, such as *Micrococcus*. There were also a few colonies of *E. coli*.

FARTING NURSE ANSWER

Luke Tennent interpreted these results as follows:

- At the moment of expulsion of the fart (called Brown Zero), the erupting gas picked up bacteria from the rectum. The gas, in a direct and powerful stream, dumped these gut bacteria onto the blood agar plate in the IBZ.
- The two incomplete arcs making up the SR imply that the outer portions of the erupting gas ran across the two inner buttock cheeks, picking up normal skin bacteria.

There is obviously a need for follow-up research after this tentative initial study. We do have one definite and real result, though: never fart while naked, at close range, near food.

Due to the embarassment factor, there is little research on farts. However, Dr Levitt, "The King of Gas", is hoping for a new medical speciality, flatology ... pointing out that there have been several cases where analysis of the flatus gas has led to the patient's diagnosis.

However, perhaps the operating theatre clothing of the surgical nurse would have had a filtering effect on the erupting gas. More layers would mean more filtering, and if the "Toot Trapper" can capture tiny molecules of gas, it could certainly capture much larger bacterial cells. And the anaesthesiologist who probably infected patients by farting may not have been wearing any underwear underneath the theatre clothing.

But these are all assumptions. Until we do more studies with blasts through clothes, the real answer will be, as Bob Dylan's song says, "Blowin' in the Wind".

References

Professor Terry Bolin and Rosemary Stanton, *Wind Breaks: Coming to Terms with Wind*, Allen & Unwin, Sydney, 1997.

Stephen S. Hall, "Deflating beans", *Science 84*, July–August 1984, pp 87–88.

F.L. Suarez, J. Springfield and M.D. Levitt, "Identification of gases responsible for the odour of human flatus and evaluation of a device purported to reduce this odour", *Gut*, Vol. 43, No. 1, July 1998, pp 100–104.

Gail Vines, "Gone with the wind", *New Scientist*, No. 2127, 28 March 1998, p 59.

Stephen Young, "A Christmas digest", *New Scientist*, 19–26 December 1985, pp 24–27.

THIS BOOK IN 60 SECONDS

The chicken that had no haid also could not lay an egg. So instead he cooked some French toast and piled on the chilli. The chilli was so hot that he bolted out of the yard and ran straight over hot BBQ coals without even noticing. Still running, he soon found himself on the Great Wall of China. Waving frantically, he couldn't raise the attention of the astronauts on the International Space Station, nor that of the Loch Ness monster (there could be a reason for that). This dissin' shocked him so greatly that his hair turned white and his little chicken belly button instantly filled with blue lint. With all this excitement, he really needed a drink. Unfortunately, he had a few too many, and the chicken coop really rocked and rolled around that night. When he awoke the next day, the coop really stank. The smell reminded him of the time when he ate too much grain, and blew the roof off the coop when he "let one rip". The next day, he headed down to the bank to get some money so he could rebuild the coop roof. The queue at the bank was pretty long, so he went outside to use the ATM. As he glanced across the road he sensed the even longer queue outside the women's toilets and thanked the Chicken Lord that he wasn't a hen.

new MOMENTS IN SCIENCE #1

"Pigeons landed a couple of scientists a Nobel Prize, because somebody bothered to discover the difference between Pigeon Poo and the Origins of the Universe."

Proving that fact is stranger than fiction, Dr Karl Kruszelnicki launched his *New Moments in Science* series with this fabulous collection of stories about some recent awesome discoveries in science.

How can sexual intercourse spoil your whole day (because you can't remember any of it)? What really killed Elvis Aron Presley (and why was he buried twice)? Why do men have nipples? And why can the Pill make a woman choose Mr Wrong?

Dr Karl gets to the bottom of them all — and more — and then goes on to crack the big one: how *did* pigeon poo unlock the secret of the universe?

new MOMENTS IN SCIENCE #2

Yet more bizarre but true stories about the latest discoveries in science, by Australia's livewire guru of science, Dr Karl Kruszelnicki.

How can a string of big, fat lasers take the place of anti-missile missiles? What are engineers doing with a tuna called Charlie the Robofish? Do you know that bacteria live in strange, slimy skyscraper cities, and that some bacteria have a tiny electric motor?

Which star, 300 000 years ago, blew out a hole in our galaxy? Is it true that beer has fewer calories than skim milk? What should you do if you are trapped in quicksand?

Do you know that cordless phones can cause strokes? And that really soon (any time within the next few thousand years) we should flip into the next Ice Age?

new MOMENTS IN SCIENCE #3

There's an old saying that scientists have, "It's not the answer that gets you the Nobel Prize, it's the question." New Moments in Science #3 has lots of questions — and quite a few answers too.

If you're caught in the rain, should you walk or run? Which wine goes best with that piece of flattened fauna? How does a cool bath help you win a marathon, and survive winter better? Is there life elsewhere in our Solar System? Will we ever be able to send people to the stars — and back to Earth again? What is the story behind Noah's Flood? Why are people called Smith heavier than people called Taylor? How can maggots that eat your flesh be good for you? And are you a supertaster?

new MOMENTS IN SCIENCE #4

"A house is built of stones, and science is built of facts. But a house is more than a pile of stones, and science is more than a bunch of facts."

How does fidgeting help you lose weight? How do you use explosives to tenderise meat and save $14 million a year? Do your ears really grow bigger as you get older? If glass is stronger than steel, why don't we use it to make buildings? Which animal is the fastest eater? Did the military invent the Internet? Why is every movie star related to Kevin Bacon, and how can this make your Internet connection faster? Did a guy called Murphy really come up with Murphy's Law? Why does toast always land buttered side down? Why are your suicidal cells essential for life? Does Therapeutic Touch really heal? And wouldn't you be unlucky if you were allergic to sex and water . . .